THE BLUE STRAGGLER MYSTERY

Other World Scientific Titles by the Author

Mind the Gap: The Labyrinthine Story of Planetary Orbits, Mathematics, and the Titius-Bode Rule
ISBN: 978-981-12-7397-1

A Cabinet of Curiosities: The Myth, Magic and Measure of Meteorites
ISBN: 978-981-12-2491-1

Going Underground: The Science and History of Falling through the Earth
ISBN: 978-981-327-903-2

THE BLUE STRAGGLER MYSTERY

Martin Beech
University of Regina, Canada

World Scientific

NEW JERSEY • LONDON • SINGAPORE • BEIJING • SHANGHAI • HONG KONG • TAIPEI • CHENNAI • TOKYO

Published by

World Scientific Publishing Co. Pte. Ltd.
5 Toh Tuck Link, Singapore 596224
USA office: 27 Warren Street, Suite 401-402, Hackensack, NJ 07601
UK office: 57 Shelton Street, Covent Garden, London WC2H 9HE

Library of Congress Control Number: 2025942252

British Library Cataloguing-in-Publication Data
A catalogue record for this book is available from the British Library.

THE BLUE STRAGGLER MYSTERY

ISBN 978-981-98-2008-5 (hardcover)
ISBN 978-981-98-2009-2 (ebook for institutions)
ISBN 978-981-98-2010-8 (ebook for individuals)

For any available supplementary material, please visit
https://www.worldscientific.com/worldscibooks/10.1142/14494#t=suppl

Desk Editor: Amalina Lai

Acknowledgements

"Let me recommend this book — one of the most remarkable ever penned," was how Sherlock Holmes, in *The Sign of Four*, introduced his faithful companion, John Watson, to William Winwood Reade's *Martyrdom of Man* (published in 1872). While every author believes that their book is the best that could be written, every author will, indeed, must, acknowledge that their work is the result of many contributors. These contributors may not have pushed the pen (or pressed a key pad), but they quietly influenced and shaped every part of the production. Indeed, all books are the result of many forces coming together at the right time, and in the right place. Some of the necessary inspirations to write a book are subtle and are hardly noticed by the author until matters are well advanced. Other inspirations are more obvious, and this author begins by acknowledging a debt of gratitude to the faculty members at the Astronomy Department at the University of Sussex in the mid-1970s. I definitely owe a debt of gratitude to Gael Ogilvie who, long ago in Edmonton, Canada made me a blue straggler birthday cake. I additionally owe a great debt to the faculty members in the Department of Physics and Astronomy at Western University, as well as to colleagues at Campion College, and the Department of Physics at the University of Regina. Many thanks are also extended to Dr. Aaron Geller for allowing the use of his image of a blue straggler under formation to grace the cover of this book. Most of all, however, I owe a great debt of gratitude to Carmen Teo Bin Jie and Nur Amalina Lai Shan Shan, at World Scientific, for all their editorial work and numerous helpful suggestions.

Contents

Introduction

There should be no combination of events for which the wit of man cannot conceive an explanation.

(*The Valley of Fear*)

Astronomers love nothing better than a mystery, and, fortunately, the Universe has never failed to provide an abundance of intrigue. It is the mysterious that captivates the imagination, and it is the mysterious that marks the boundary between the known and the unknown. The mysterious provides the exception to the rule, and in this manner the blue straggler mystery is set against the backdrop that surrounds the first detailed phase of calculations relating to stellar evolution. While the first stars that turned out to be called blue stragglers were noted as anomalous in the early to mid-1950s, it was not until the early 1960s that they were clearly identified and accepted as being a distinct and *bona fide* group of objects. Prior to this time, throughout the vast tailback in the long history of astronomy, their existence was never expected nor predicted. Mysterious in name, and mysterious in place, the blue stragglers presented astronomers with a challenge; they were misfits, and in rising to the challenge of identifying them, astronomers were eventually forced to contemplate the most extreme circumstances of stellar interaction and evolution. The blue stragglers bucked the trend of their siblings: they are not your typical star, and remarkably, unlike a myriad of human alchemists, they have succeeded in finding an elixir for rejuvenation (although the process of growing young, as will be seen, requires the sacrifice of another star).

"Data! Data! Data! … I can't make bricks without clay," declared Sherlock Holmes in *The Adventure of the Copper Beeches* — and philosophically he was absolutely right. In order to know something, one needs data, and in the sciences data is derived by measuring something. With respect to the blue straggler mystery, measurements of temperature and brightness for many hundreds, even thousands, of stars in the same star cluster were needed (how this is done is the topic of Chapter 1). Indeed, by looking at the group properties of stars within a given star cluster, it is possible to find out when the cluster actually formed, and it is with respect to these measurements that the blue stragglers stand out. As Chapter 2 will reveal, the evolution of a star is largely governed by its initial mass, and the more massive a star, the more rapidly it changes with time. The internal and external changes of a star are measured by size, surface temperature, and luminosity. With respect to all of these metrics, blue stragglers have apparently evolved differently. In 1994, Yale astronomer Charles Bailyn commented on what he considered to be one of the first triumphs of the Hubble Space Telescope (launched in 1990): the discovery of some 21 blue stragglers in the core of a nearby globular cluster called 47 Tucanae. Certainly, this was an unprecedented and important observation, but more strikingly, Bailyn commented that, "blue stragglers have been a thorn in the side of students of stellar evolution since their discovery 40 years ago" [1]. Even now, astrophysicists struggle to fully explain the blue straggler phenomenon. Certainly, we know much more about blue stragglers today, but they remain mysterious and enigmatic.

Stellar structure and stellar evolution is the grist to the astrophysics mill, and they describe how a star is put together, how they work, and why they change with time. Indeed, astrophysics is a form of reverse astrology, and through its application human beings tell the proverbial stars what's in their past, present, and future. Specifically, for our story, what makes the blue stragglers stand out from their companions is that they have luminosities and temperatures that are higher than would be expected from the deduced cluster age. The blue stragglers appear to straggle in their evolutionary progress relative to the other stars of the same mass in their host clusters, and, of course, the question is "why?". Indeed, this is the mystery of blue stragglers. Why, astronomers ask, do a handful of stars out of the many hundreds to many thousands of stars within a cluster evolve in a different, seemingly slower way to their same mass cohorts? What physical process has seemingly re-set or slowed their evolutionary clock, making them appear younger than their siblings? Blue stragglers apparently do not follow the expected trends dictated by the standard theory of stellar evolution, and this tells us that something else must be going on — something that astronomers had not included in their stellar models. Sherlock Holmes summed up this kind of discrepancy (in a reverse sense) in *The Sign of Four*, where he noted that, "while the individual man is an insoluble puzzle, in the aggregate he becomes a mathematical certainty."[1] For us, the blue straggler is the individual, while the aggregate is the cluster. Fortunately for us, the individual (the blue straggler) does not represent an unsolvable puzzle. That is not to say that astronomers know all there is to know about these wayward stars. As pioneering astrophysicist Sir Arthur Eddington once remarked (in a different context[2]), "something unknown is doing we don't know what." The quest of the astronomer has been to find out what the 'unknown' is, and what exactly the 'unknown' is doing.

So, in summary, at this stage, all we have acknowledged is that blue stragglers are a real phenomenon. They appeared on the *scene* at about the same time that the Beatles gave us a yellow submarine and the blue meanies, yet astronomers were unsure as to their origin. Observationally, they are a small subgroup of stars, set within the vast assemblages of star clusters (see Chapter 1). They can be observed within both relatively young open clusters and the oldest

[1]This comment is taken directly from Winwood Reade's *The Martyrdom of Man* (1872). Reade further adds, however, "as an individual, he is a free agent: as a species, the offspring of necessity." In grand Holmesian fashion, Reade also noted that, "life is one long tragedy; creation is one great crime."

[2]The context was that of how stars generate energy by fusion reactions within their central cores (see chapter 2).

Figure 1. The globular cluster Messier 3. Of the 500,000 stars within M 3, some 100 have (so far) been identified as blue stragglers. The cluster is located about 10,000 parsecs away from the Sun, and it is about 60 parsecs across. Image courtesy of HST/NASA.

globular clusters (Figure 1); they can be found at the center of our Milky Way Galaxy, and within nearby dwarf galaxies. Indeed, blue stragglers are found wherever stars tend to cluster together, and this is an important clue to their origins.

The blue stragglers appear younger than their same-mass counterparts: they are hotter (what astronomers call bluer — see Chapter 1) and more luminous. The stragglers are seemingly either formation laggards, or they have discovered some special cosmic alchemy to extend or regain their youth. In some sense, an obvious solution to the straggling problem is to posit that star clusters are composed of many stars formed at significantly different epochs — that is, the cluster stars did not all form at the same time. In this manner the blue stragglers are formation stragglers, and are simply ordinary stars that appeared at a later moment than the other stars within the cluster. In this sense, the vast majority of cluster stars are simply older than the blue stragglers. The argument of delayed star formation fits the facts (as presented so far), but it transpires (see Chapter 3) that there are very good arguments to indicate that this is not the answer to the blue straggler mystery. Indeed, stranger and more extreme ideas will need to be considered before the blue stragglers can be put into their appropriate astronomical pigeonhole.

Blue stragglers, as revealed earlier, are a relatively recent discovery. And coincidently, they also appeared at a time when astrophysics was entering a golden era. Indeed, following the ascendency of astrophysics in the first half of the 20th century, theorists in the mid-1960s were beginning to get a good handle on how stars with different masses change (that is, evolve) as they age. Indeed, the 1960s was the epoch of the first detailed (electronic) computer models, and the key requirement at that time was the calibration of parameterized model terms. To do this it was necessary to look at stars with the same age and composition, but different initial masses, and the ideal testing/calibration ground was found within the Hertzsprung—Russell (HR) diagrams of star clusters (see Chapter 1). In turn, these diagrams needed to be produced to a high standard of precision, with extensive catalogs of star characteristics being constructed. The key idea behind such comparisons was that all the stars within a given cluster formed at the same time, and from the same interstellar cloud. In this manner the start of the star formation clock is synchronized, the individual star compositions are the same, and the only variable term left is that of the individual initial masses appropriate to each star. It is additionally important to consider whether a specific star has a close binary companion — but we will come to this later. Once an HR diagram has been constructed, various computer model parameters can be adjusted to bring the theoretical models into agreement with the observations. By this give-and-take approach to adjusting the numerical model parameters, astronomers began to increase their understanding of just how a typical star works.

Figure 2 shows the history of academic research papers relating to blue stragglers. The data is taken from the Astrophysics Data System (ADS)[3] and covers those articles found within professional journals and astronomy magazines since 1963. Although the first

---

[3] The database can be accessed at https://ui.adsabs.harvard.edu

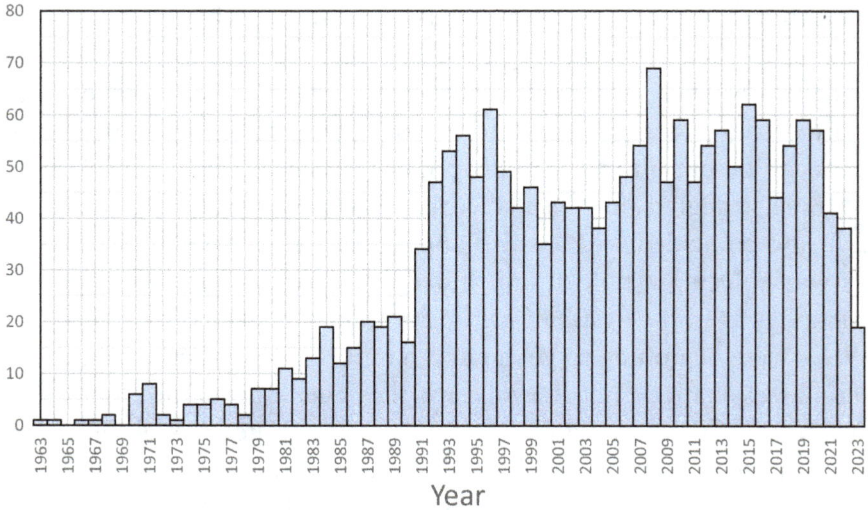

Figure 2. The number of publications per year, between 1963 and 2023, where the term "blue stragglers" appears in the abstract and/or the title field of a published research paper. Data from NASA's Astrophysics Database.

observations of what are now called blue straggler stars were made in the early 1950s, the name blue straggler was only coined later. As can be seen from the numbers, the blue straggler story essentially began, albeit slowly, in the mid- to late-1960s, with just a few research articles appearing per year at that time. The citation in which the ADS finds the first use of term "blue straggler" in the abstract of a research article is the doctoral thesis by (now) Professor Iwan Williams in 1963. William's supervisor was Sir Willam McCrea, and his thesis, presented to the University of London, Royal Holloway College, was concerned with *Processes of Evolution in Astrophysics*. The term blue straggler, had actually been coined earlier in a textual discussion (see Chapter 2), but Williams's article is the first formal usage of the term in a *bona fide* object sense. In 1979, however, J. Craig Wheeler (University of Texas, at Austin) wrote, "blue stragglers remain one of the unexplained oddities of astronomical lore" [2].

In spite of a slow and uncertain start, it is evident in Figure 2 that as we move into the late 1970s, through the 1980s, and into the

1990s, there was a steady increase in research interest and output. By the early 1990s, some 50 articles on blue stragglers were being published per year. As will be seen in subsequent chapters, the 1990s was an important decade for unraveling, what remains to this day, the essential take on blue straggler origins. Even so, writing in 1992, Virgina Trimble (University of Maryland) was still able to correctly surmise that, "all in all, blue stragglers seem to be both inadequately understood and insufficiently appreciated" [3]. These same sentiments were echoed by Lynda Stryker (Arizona State University) a year later, in 1993, when she commented that, "progress has been made in the last 40 years, but blue stragglers remain an intriguing challenge" [4].

Rarely are mysteries solved in a single step, or by invoking a single specific physical process. This certainly appears to be the situation with respect to blue stragglers. There are, as it turns out (see Chapters 3 and 4), many different ways of potentially regenerating and/or reconfiguring an ordinary star into a blue straggler. Furthermore, nestled amongst the blue straggler population of stars may be a few stars that present us with an even deeper mystery than that of stellar rejuvenation. Perhaps, some blue stragglers may be stars astero-engineered by advanced alien civilizations. This latter speculation, along with others, will be examined in Chapter 4.

Before moving on to Chapter 1, a few personal comments might not go amiss with respect to my engagement with the blue straggler story. I was fortunate, indeed, very fortunate to begin my astronomy career (such as it has been) at the University of Sussex in the early 1980s. It was the time of the Three Mile Island nuclear accident, the first appearance of the Sony Walkman, and the Rubik's cube was the latest craze. It was the time in which the first Voyager I spacecraft images of planet Jupiter were taken and received back on Earth; Mount St. Helens exploded in Washington State; Margaret Thatcher was first elected British Prime Minster; the first test-tube baby was born; the first Garfield comic strip was published; the first DeLorean cars were being manufactured in Ireland; and Pink Floyd released their now classic album *The Wall*. Like every epoch in human history, "it was the best of times, it was the worst of times."

As now, so then, the Astronomy Department at the University of Sussex was a vibrant hub of research activity. During my time there, enrolled in the Master's degree program, the main research staff included Roger Taylor, Leon Mestel, Robert Connon Smith, and (Sir) William McCrea (all giants in their fields of study) [5]. It was Roger Taylor, and Robert Connon Smith who particularly inspired my thoughts, and they set me upon a path that eventually weaved its way to a doctoral dissertation concerned with the formation and evolution of massive stars. At that time, however, I was fascinated by Wolf—Rayet (WR) stars [6], and it was post-doctoral fellow, Dr. Herb Falk who supervised my Master's Thesis on this topic. Herb was a computer expert, then a relatively new label for a theoretical astronomer to adopt, and he had just completed his doctoral thesis, in 1979, at the University of Western Ontario in Canada, on the evolution of massive stars. It was a time when computers were objects of mystery. They were obscure, large, and housed in some far-off building. Indeed, one took one's FORTRAN program down to the computer room as a stack of punched cards, hoping that there were no punch code errors, since that could cost a half-day's worth of editing time to correct.

Apart from providing me the springboard for further adventures in astronomy [7], Sussex is also home to a particular memory that I still vividly recall. It was an otherwise ordinary Friday afternoon, but Professor McCrea had decided to tidy his office of pre-prints, and journal off-prints. This resulted in the appearance, of a number of large boxes stuffed to overflowing with research articles in a small storage room. The news spread through the Department like wildfire, and a scavenging hoard of graduate students beat their way to the storage room with the aim of raiding the boxes full of treasure. The game was afoot. A thrill of discovery drove the frantic scavenging, and it was a sheer joy to be finding articles on new and exciting research topics. Before the internet, pre-prints and off-prints were the graduate student's equivalent to sports trading cards. One collected them, one treasured them — they were occasionally given away but more often exchanged by trading. The material was our source and gateway to new research and innovative ideas. For me, the key find that triggered intense interest was a pre-print publication of John Craig Wheeler's now seminal article on blue stragglers [2] — I still have this reprint, nearly 50 years later. Indeed, it was in this pre-print that I first came to know of blue stragglers, and the article itself contained headings that set my imagination racing: "delayed star formation" and "quasi-homogeneous evolution" — these were words and ideas to marvel at. What was going on, what were these stars, and why wasn't everyone talking about them? I was to find out later that it was William McCrea, [8] and his then-doctoral student Iwan P. Williams, in the early- to mid-1960s, who had first set about trying to understand how blue stragglers might actually form. The rest, as the time-honored saying goes, is history (and the contents of Chapters 2 and 3). But it is a fascinating history, with the mystery of the blue stragglers still only partially solved to this very day. Much has been learned over the past half-century about these enigmatic stars, of course, but there is still much to investigate. Like all good mysteries, that of the blue stragglers has turned-up surprises, and some answers, but has also generated more questions than can at present be fully answered.

The great detective Sherlock Holmes, or perhaps I should say his trusted companion and biographer Dr. John Watson, provided the partial inspiration behind the title to this book, and indeed, it turns out that, with only very little literary forcing, much Holmesian logic and inference underpins the narrative presented in subsequent chapters. While not the most reliable of logicians, there are many maxims from Holmes that annotate not only the story of the blue stragglers, but the ways of *doing* science in general. Accordingly, it seems worth our while to take a brief look at his methods and outlook, and to use what we find as a guide to our own thinking.

A mystical figure in his own right, Holmes is a fictional character[4] created by Arthur Conan Doyle in 1887. For all his nonreality, however, Holmes continues to evoke the very image of a larger-than-life

---

[4] Some aficionados would probably delete the word 'fictional'.

detective, albeit a high maintenance one, that continues to resonate with the reading (as well as television and film viewing) public. His grip upon the human imagination has stood the test of time for nearly one and a half centuries, and the energy and passion that he throws into his investigations are models for for the tenacity and drive needed for conducting scientific research. In his first adventure, *A Study in Scarlet*, Watson (Sherlock's trusted Boswell) tells us that Holmes has zero knowledge of astronomy, and no knowledge of philosophy. However, while he never specifically needed a detailed (insider or elementary) knowledge of these topics to solve any specific crime, it is clear that he thought deeply about how science works, and how philosophy directs scientific investigation. His philosophical quips are manifest throughout the Canon, and syllogisms of one form or another continuously trip off his tongue. For all this, there are some technical issues that deserve further investigation. Indeed, for example, perhaps Holmes's most famous (and oft repeated) maxim (found, for example, in *The Adventure of the Beryl Coronet* [9]), is one of his least trustworthy of methodologies. Here it is: "When you have excluded the impossible, whatever remains, however improbable, must be the truth" [10]. Like many similar such statements (not just those by Holmes), what seems like sound and clever logic is, in fact, upon more thoughtful analysis, a rather poor way to investigate a mystery. How can one ever truly know if a given scenario is certainly impossible? How are probabilities assigned, who made the list of scenarios, upon what criteria were they selected, and, more importantly, why must the truth automatically result from a series of knockdowns and exclusions? Sadly, Holmes's logical, and probabilistic approach cannot guarantee the success of any deductions and/or investigation. Indeed, Holmes readily acknowledges this fallacy to Watson in *The Adventure of the Yellow Face* — a story in which essentially all of his guesses, and conjectures turn out to be wrong.

In spite of the philosophical quibbles given above, to Holmes, a mystery was generally solvable according to a four-step process. Firstly, one has to take-in the fact that there is a mystery to be solved, and this is essentially a listening and reading process. What is the problem, who is involved, and what has happened to bring the mystery to anyone's (and specifically his) attention. In this manner, nearly every adventure in the Canon begins with a client relating their story to Holmes at 221B Baker Street. Either in person, or via a letter, the problem, or mystery, is described and identified. What Holmes does next is investigate. He gets his boots on, he crawls on the floor, he sniffs the empty glasses, he snatches a magnifying glass from his pocket, and he picks up cigarette ash, he measures the stride length between muddy footprints, he assesses the lie of the land, and he gauges the depth of bicycle tyre tracks. That is, he gathers physical data that he can work with. And by such activities, he generates nuggets of solid and useful information about the mystery at hand: a timeline of events, and an understanding of the *dramatis personae*, the means of their locomotion, and their habits. Stage three is where Sherlock applies his highly trained mind to the data (the hard-won facts) that have been unearthed during the observational exertions of stage two. Here, whether presented with a "three-pipe problem", or a trivial set of circumstances, Holmes turns his mind to the generation of scenarios that explain the known facts (figure 3). It is in this manner that he applies logical reasoning to unravel the possible chains of circumstances and events that can account for an understanding of the mystery, and from which he identifies a most likely answer. In this way he moves to stage four and the denouement (the trapping of the villain, or the explanation of some mystery object or symbol). At this final stage, loose ends are brought together and all is explained to an amazed audience — including the importance of seemingly trivial threads that had been dismissed by everyone else. The astronomer is not quite so fortunate as Sherlock Holmes, in that while stages one, two and three are certainly part of the scientific process, the final explanatory stage, stage four, is rarely the end of the story. Science is typically more provisional in its stage four answers, and while the major culprits and protagonists might well be identified, there are often gaps in the chain of reasoning [11], along with additional oddities and intransigent loose ends that simply won't be tidied away. Indeed, in *The Sussex Vampire*, Holmes reminds us that, "one forms provisional theories and waits for time

Figure 3. Sherlock Holmes in contemplative mood. From *The Man with the Twisted Lip*, in which Inspector Bradstreet says, "I wish I knew how you reach your results." Artwork by Sidney Paget [12].

or further knowledge to explode them." He also acknowledges that the available data can often be interpreted in many different ways: "I have devised seven separate explanations, each of which would cover the facts as far as we know them," he tells Watson at an early stage in *The Adventure of the Copper Beeches*. By adopting the tools contained within Holmes' philosophical carpetbag, however, astronomers, or any scientist for that matter, won't go too far astray in finding at least provisional answers to any of the questions that might be posed by the marvelous Universe around us.

> *One's ideas must be as broad as Nature if they are to interpret Nature.*
>
> (*A Study in Scarlet*)

Notes and references

[1] C. Bailyn. The case of the blue stragglers. *Science*, **263**, 44–45, 1994.

[2] J. C. Wheeler. Blue stragglers as long-lived stars. *Astrophysical Journal*, **234**, 569–578, 1979.

[3] V. Trimble. Astrophysics in 1992. *Publications of the Astronomical Society of the Pacific*, 105, 1–21, 1993.

[4] L. Stryker. Blue Stragglers. *Publications of the Astronomical Society of the Pacific*, **105**, 1081–1100, 1993.

[5] https://www.sussex.ac.uk/research/centres/astronomy/people/history.

[6] First described by Charles Wolf and Georges Rayet in 1867, these stars were initially a mystery to astronomers due to the fact that they exhibited strong emission lines superimposed upon an essentially continuous spectrum (see Chapter 1). The majority of stars, in contrast, show only absorption lines in their spectrum. It was eventually realized that the emission lines in WR stars were due to highly ionized carbon, nitrogen, and oxygen atoms, and were associated with strong stellar winds. Ultimately it was found that WR stars are associated with a late-stage evolutionary phase of massive stars.

[7] Following Herb Falk, I eventually completed a doctorate at his old *Alma Mater*, the University of Western Ontario (now Western University), on *The Formation of Massive Stars: a 30 $M_\odot$ case study*, in the early 1990s. I even had the same supervisor as Herb for this work: Professor Romas Mitalas. Indeed, it was with Romas that I first investigated the formation and evolution of WR stars, along with the characteristics of massive blue stragglers, and supermassive, chemical homogenous stars. From Western, in the mid-1990s, I joined the faculty of Campion College, at the University of Regina, and taught astronomy there for many years.

[8] W. H. McCrea. Extended main-sequence of some stellar clusters. *Monthly Notices of the Royal Astronomical Society*, **128**, 147–155, 1964.

[9] Similar such arrangements of this Holmesian saying are found in *The Sign of Four*, *Silver Blaze*, *The Adventure of the Blanched Soldier*, and *The Adventure of the Priory School*. Clearly Doyle thought it a sound philosophical principle upon which to investigate crimes.

[10] Quotations, in all chapters, are taken from the extensive Canon of adventures relating to Sherlock Holmes. These include: *A Study in*

Scarlet (1887), *The Sign of Four* (1890), *The Adventures of Sherlock Holmes* (1892), *The Memoirs of Sherlock Holmes* (1894), *The Hound of the Baskervilles* (1901), *The Return of Sherlock Holmes* (1905), *The Valley of Fear* (1914), *His Last Bow* (1917), and *The Case-Book of Sherlock Holmes* (1927). Additional quotations are taken from Doyle's stories featuring Professor Challenger (the more-blustery and tyrannical *alter ego* of Holmes) — *The Lost World* (1912), *The Poison Belt* (1913), *When the World Screamed* (1928), and *The Disintegration Machine* (1929). Quotations have also been adopted from *The Exploits of Sherlock Holmes* (1954) by Adrian Conan Doyle (A. C. Doyle's youngest son) and J. D. Carr. There are numerous books relating to the adventures and exploits of Sherlock Holmes. A good text for the new reader to begin with, however, is Michael Harwick's, *The Complete Guide to Sherlock Holmes* (St. Martin's Press, New York, 1986).

[11] In *The Musgrave Ritual*, one of the key steps in Holmes's solution is only available by luck (or, an engineered turn of literary good fortune). Without the fortunate prior calculations made by Reginald Musgrave, many years earlier, the discovery of the starting point for the rhyming sequence would not have been possible. Indeed, this point serves to remind us that scientific progress is also made, on occasion, by the good grace of *Fortuna*. As with the treasure hunter, the astronomer is more than happy to make progress on the back of a lucky break — just ask any amateur comet sweeper. It is also in the *Musgrave Ritual* that Holmes makes the comment that in order to unravel the meaning of the ritual he made, "allowance for the personal equation, as the astronomers have dubbed it", indicating that Watson was wrong, when in *A Study in Scarlet* he claimed Holmes knew nothing of astronomy.

[12] Image from: https://commons.wikimedia.org/wiki/File:Sherlock_Holmes_-_The_Man_with_the_Twisted_Lip.jpg.

Chapter 1

The measure of the stars

The game is on

(*Sherlock — TV Series: A Study in Pink*)

The stars are extremely remote from us. We cannot travel to the stars nor survey them *in situ*, so we have to rely purely upon observations that can be made from Earth, or Earth's orbit, to deduce their properties. In order to do this, astronomers have had to learn how to read the hidden messages that are woven into starlight. Indeed, astronomy, both ancient and modern, is predicated upon what we can see, rather than what we can touch and manipulate[1], and what astronomers ultimately have to work with is the data that is provided by the instruments and technology available to them in their specific lifetimes. This chapter, accordingly, is all about methodology. More than any other generation before, we live in a time when access to information is a blink of an eye away: data flies around the internet at close to the speed of light, with little reference to where it came from, or where it might be going to. We are both drowning in and spoilt by an abundance of data. We have the assembled numbers but often have little clue as to their meaning. Look into any introductory astronomy text and one will find numerous tables and diagrams

[1] The exceptions to this are those studies concerning planetary materials, meteorites, asteroid and cometary samples that can be deep-searched and studied under the microscope in a terrestrial laboratory.

showing the results of near countless observing hours by astronomers, and while the data informs us about stars, it doesn't tell us how it was gained. It behooves us now, therefore, to look into the basics of how the numbers and data points were obtained. We do not need to look at all the subtleties, approximations, limits, error analysis, or fine-tuning of the data gathering process (there are many learned tomes that cater to this function), but what I would like to show in this opening chapter is that there are straightforward means by which the fundamental physical properties of very distant stars can be gleaned without leaving the surface of the Earth. This chapter is all about the development of an astronomical toolbox; a toolbox full of practical observational tools and theoretical means, with which human beings can explore and indeed, understand (at least some of) the vast Universe about us.

To the astrophysicists, the romance of astronomy is only veneer-thin. What really counts, in spite of any reproachment from the ghost of Walt Whitman[2], are the numbers arrayed and the charts full drawn. To understand the stars' enumerated data is vital. Indeed, without knowledge of their physical characteristics in number form, we know essentially nothing about what they really are. Historically, astronomy concentrated on just two stellar attributes: the location of a star on the sky with respect to other stars, and the brightness of that star relative to those about it. Indeed, that's about all that could be gauged of star properties until well into the 19th century. The invention of the telescope in the early 17th century did little, at that time, to enhance our knowledge of what the stars actually are. Certainly, early telescopes revealed more stars to the eye, but these newly found stars looked no different to the ones that could be seen with the naked-eye — they were just fainter. To move beyond the superstitions of astrologers, and the euphuistic musings of the poet (all well and good in understanding the human lot), the astronomer first had to learn how to organize and collate their observations, and then learn how to interpret the numbers and columns derived from

[2] See Whitman's oft' quoted poem: *When I heard the Learn'd Astronomer*.

those observations. This is a process that has taken many centuries to achieve, and indeed, it is still a vibrant, and ongoing activity.

In its reduced, rendered, codified, and calibrated scientific form the properties of an individual star can be expressed as a n-tuple of the form,

$$\text{Star}(k) = (Q, V, N, D, E, X, T, L, R, M, A),$$

where $k = 1, 2, 3, \ldots,$ identifies the individual star[3], and where the meanings attached to the various letters are:

Q: Location on the sky. Technically this must be a set of two numbers providing the star's declination and right ascension. This position identifies the constellation in which the star is found and fixes the star's sky position, at some specific epoch, relative to its neighbors. The position of the star in the sky (both above or below the horizon) to an observer in any location on earth, at any chosen time, can also be calculated from Q.

V: Motion along the sky. Once again, this is technically given as two numbers relating to the yearly rate of proper motion in declination and right ascension. Importantly, combining proper motion data with parallax (that is, distance D) and Doppler velocity (V_R) information enables the determination of the star's actual velocity through space.

N: System number. $N = 1$ corresponds to a single star, $N = 2$ indicates that the star is part of a binary system, $N = 3$ indicates that the star is part of a triple system, and so on.

[3] Any given Star(k) will typically have many catalog names and identification numbers. The Pole Star, for example, is concomitantly known as Polaris, α Ursa Minoris A, HD 8890, SAO 308, and HIP 11767 (amongst other catalog designations). Star data and aliases can generally be found in SINBAD (the set of identifications, measures and bibliography for astronomical data), which is maintained by Centre de Données astronomiques de Strasbourg (CDS).

D: Distance to the star from the Sun. The distance to a star can be determined by the use of parallax observations, or through a series of calibration steps.

E: Environment. This designation will become more meaningful as we proceed, but the broad idea is to locate the star as being either a field star, i.e., a star relatively isolated from other stars, or a member of a galactic or globular cluster. For us, this will largely determine the applicable number density of stars per unit volume of space.

X: Composition. The make-up of the star is revealed through its absorption spectrum and specifically the absorption lines visible in the spectrum.

T: Temperature. The temperature is again determined by an analysis of the absorption line features that can be seen in the spectrum of the star. The temperature can also be gauged from the wavelength at which the star radiates the greatest amount of electromagnetic energy into space — this follows from Wien's displacement law (see below).

L: Luminosity. This is a measure of the energy radiated into space (in all directions) per second over all wavelengths of electromagnetic radiation. It cannot be measured directly, but can be determined by measuring the energy flux, *f*, through wavelength-specific filters, and calibrating these measures against the flux expected from a black-body radiator of temperature *T*.

R: Radius. This can be measured directly in the case of eclipsing binary star systems, or is inferred from the deduced temperature and luminosity of a star (and an application of the Stefan–Boltzmann law — see below). In a very few cases, the size of a single star can be determined if its distance (*D*) and angular size are known.

M: Mass. This can only be measured directly for stars of binary systems. Essentially, it is based upon the separation of the stars, i.e., the distance of each star's center from the system's center of mass, and a measure of the orbital period. This result is based upon an application of Kepler's 3rd law.

A: Age. This is not a characteristic that can be measured directly. It is inferred by placing the star within some evolutionary context as determined by its measured quantities *X*, *T*, *L*, *R*, and *M*. In the case of the Sun, its age can be directly inferred from the measured ages of meteorites in the laboratory — giving an age of 4.56 billion years. For all other stars, age estimates are determined by comparing their physical characteristics against evolutionary stellar models.

Having introduced our Star(*k*) *n*-tuple[4], along with its various terms, it is now necessary to see how the specific entries are deduced and codified. Indeed, it is this latter action that cuts to the very core of what modern astronomy and astrophysics is all about. These are the same actions that provide data from which solid (well, arguably firm) conclusions can be drawn[5].

1.1. Charting the heavens

Every one of his actions was directed towards a definite end
(*Boscombe Valley Mystery*)

Observational astronomy is perhaps the second oldest profession. Before written records of any form were even being kept, the heavens were under human scrutiny. Since the most ancient of times,

[4] In our case $n = 11$, but larger values of n, introducing other measured quantities, can certainly be envisioned. Effectively, however, each n-tuple can be thought of as a line from an extensive database of star properties and observations.

[5] As noted in the Introduction, however, all observations are theory-laden, and data is almost invariably gathered under one theoretical assumption or another. In a very real sense, no one ever sets out to find anomalous data — it is something that happens, by accident or luck, when looking for something else. The data gathering process invariably has a specific aim to test one idea or another. If the data, so deduced, does not fit in with respect to some favored theoretical framework, then this is an indication that a new theoretical outlook is required.

someone, somewhere, was likely observing and reacting to nighttime conditions and events. What was the phase of the Moon? Which constellations were visible just after sunset or before sunrise? At what point on the horizon did the Sun set, and at what point on the horizon did it rise? Were temporary and/or unexpected phenomena observed last night: a lunar eclipse, a comet, a meteor, or a new star? And so on. When it comes to the heavens, there is regularity to be recorded and random novelty to be unsure about. One thing, however, was clear very early on in the development of astronomy, and that was the simple observation that the stars do not appear to move relative to each other. Yes, they move across the sky during the night, and some stars are only seen above the horizon in certain seasons, but they are always in the same relative configuration, year in, year out. This apparent relative stability of star positions, however, is just an illusion, and in fact, the stars do move relative to each other on the sky (and in space) — they just do it, from our human perspective, very, very slowly. The typical space velocity of stars as they move around the galaxy (in the Sun's neighborhood) is in the order of 150 km/s, though some are faster, and some are slower. By human standards, this is fast, but given the vastness of space, and the large distances between us and the stars, the time required to move through even a very small relative angle on the sky is typically of order centuries to millennia.

That stars really do move relative to each other was first deduced by Edmund Halley in 1718. Indeed, it was a conclusion based on incredulity. Comparing the recorded positions for Sirius, Procyon, and Arcturus (three of the brightest stars in the sky) as recorded in both Ptolemy's *Almagest* (produced circa the first century AD) and contemporary star charts, Halley realized that the three stars had appreciably shifted with respect to their neighbors. Indeed, the amount of shift was sufficiently large that he could not accept that it was an issue of either observational error or translational slip. The brightest, most obvious stars in the sky moved relative to the fainter background stars, and this suggested that all stars actually move. The so-called fixed stars were not actually fixed at all, but moved through space relative to each other. This sky-shift is measured in terms of what is called proper motion, μ, and is given as some angular shift per year. The proper motion for Sirius amounts to $\mu = 1.3$ arcseconds per year; that for Arcturus is $\mu = 2.3$ arcseconds per year. Barnard's star has the greatest recorded proper motion (at the present time), with $\mu = 10.3$ arcseconds per year. Given enough time, the stars will shift and shuffle around the celestial sphere, scrambling and re-configuring the constellations.

The idea of constellations, marked out by a specific grouping of stars, is old. Indeed, many of our present-day constellations have configurations and names that date back many thousands of years, and were assigned by the ancient Babylonians. The notion of constellations is purely one of simplified recognition and the fashioning of a heavenly backdrop for reference. Indeed, the rise and set times and the visibility at night time of specific constellations enables an observer to gauge the time of year. The constellations also act as background markers for the position of the Sun, or the Moon, or the location of a planet. The shapes of constellations are entirely arbitrary, and are constructed according to human imposition, but they are useful sub-divisions of the celestial sphere. While their official (or generally recognized) number has varied over the centuries, the current number of constellations is set by the International Astronomical Union at 88 — with names running the alphabet from Andromeda to Vulpecula. In terms of sky area covered, at 1302.8 square degrees, Hydra is the largest constellation; at 68.4 square degrees, Crux is the smallest.

Each constellation, with varying levels of recognizable shape, is mapped-out by a specific set of stars, and the stars themselves are further classified according to their brightness. That the brightness of the stars varies from one to another is obvious to the eye, but it is conjectured that the idea of assigning a specific magnitude to star brightness began with Hipparchus in the 2nd century B.C. The magnitude scale ranks a star's brightness from 1 to 6, with 1 being the brightest and 6 being the faintest. This ranking is based upon human perception, allows for fractional values, and is related (in modern terms) to the actual flux of electromagnetic energy (over visible wavelengths) received at the Earth's surface. In modern times, the magnitude scale has been extended and made much more rigorous, but this will be discussed in more detail later.

The celestial coordinate system, *Q*, places the Earth at the center of the universe. This entirely anthropocentric approach is allowable[6] since it is only a mapping system for recording the positions of objects on the sky. The two key coordinates (Figure 1.1) are right ascension (RA) and declination (Dec. or δ). The zero point for declination is really just a circle around the entire sky corresponding to the celestial equator. An object with zero degrees of declination lies somewhere on the celestial equator. Objects northward of the celestial equator have positive declination values, while those southward of the celestial equator have negative vales. To fix the position of the object around the sky, the right ascension is measured with respect to the zero point of the vernal equinox (VE). The VE itself is located on the sky at the point where the ecliptic intercepts the celestial equator. The ecliptic, in turn, maps out the path that the Sun takes during the course of one year, around the celestial sphere. The right ascension is measured eastward along the celestial equator, and is usually cast in units of hours or degrees, with one hour of RA being equivalent to 15 degrees of motion along the celestial equator. Other

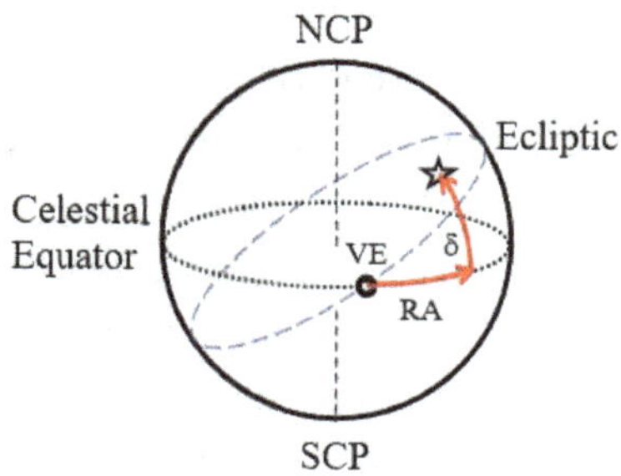

Figure 1.1. The celestial sphere and the right ascension (RA), declination (Dec. or δ) coordinate system. The Earth is taken to be at the center of the celestial sphere, and VE corresponds to the location of the Sun, on the celestial equator, at the time of the spring or vernal equinox.

---

[6]The so-called Copernican Principle is typically invoked in such a way that there must be nothing special or central about our location, as observers, on the Earth.

coordinate systems are sometimes used for star positions, but the RA and Dec. system is a general system that privileges no specific observer on the globe.

In addition to fixing the relative positions of the stars on the sky, astronomers are also interested in determining the locations appropriate to the Sun, Moon and planets, at specified times, relative to the stars. For some three-quarters of the past two-thousand years, the theory of planetary motion was that assembled and developed by Claudius Ptolemy in the 1st century A.D. The course of astronomy only began to change in the 16th century with the reintroduction[7] of a heliocentric cosmology by Copernicus that placed the Sun at the center of the planetary system. Not only did Copernicus shift the Earth away from the center of the planetary system (the then-perceived center of the Universe), other philosophers soon did away with the idea of a compact celestial sphere or shell of stars about the Earth. The stellar realm was allowed to expand to unseeable, and indeed, unfathomed depths. The appearance of the optical telescope in the early 17th century did little to change this stellar viewpoint — it simply revealed more and fainter stars. The stellar realm only began to take on its modern form, moving away from position measurements and the production of star charts, in the mid-19th century. What revolutionized the practice and direction of stellar astronomy, and indeed, ushered in the modern era of astrophysics, was the development of spectroscopy.

1.2. Spectroscopy

You are aware that no two thumb-marks are alike

(*The Norwood Builder*)

Just as a careless fingerprint left at a crime scene can be used to identify the criminal, so too can a stellar spectrum provide the fingerprint

[7]The term reintroduction applies, since Greek philosopher Aristarchus of Samos had considered a Sun-centered cosmology as early as the 3rd century BC. The great contribution by Copernicus was to develop the heliocentric concept into a workable mathematical scheme.

identity to a star. Spectroscopy (our stellar parameter X) is all about the atomic messages that are encoded in starlight. And these messages are as unique and undeniable as the fingerprints of a burglar.

The deep interior of a star is composed of a dense gas of atomic nuclei and electrons. Indeed, the conditions are such that neutral atoms, that is atoms having a nucleus (composed of protons and neutrons) and a full complement of associated electrons, cannot exist there. Pioneering astrophysicists Sir Arthur Eddington, in his 1926 book *The Internal Constitution of the Stars*, described the internal conditions within a star as being akin to a "jolly crockery-smashing stall". Every time an electron attempts to embrace an atomic nucleus, it is instantly rebuffed by a collision or interaction with a photon. There is no rest for the long-suffering electron, or photon, within the interior of a star. This fully ionized state, however, begins to change as one moves towards the outer layers of a star, where the temperature and density attain their lowest values. This region, the photosphere, is where the photons can truly begin to fly, and indeed, escape from the star into the depths of space (or into a telescope). In the deep interior, the continuous interaction of radiation with the crashing of nuclei and electrons dictates that even though a photon travels at the speed of light, it might travel just a few millimeters before its course is redirected. If left to its own devices, an unimpeded photon would travel from the center of the Sun, through the photosphere, and out into space within about 2 seconds. Remarkably, however, the continuous absorption, emission, and random encounters of a photon, dictates that it will take of order 50 thousand years to wind its way through the Sun's interior to an eventual escape in the Sun's photosphere. Importantly, however, in the cooler regions of the photosphere, atomic nuclei can begin to hold on to electrons, either in a neutral or ionized state, and under these conditions a whole new set of physical interactions with the radiation field begin to take place. It is these new absorption and re-emission interactions that allow for the development of a comprehensive diagnostic scheme for the stars. These same interactions also allow us to determine the chemical makeup of a star; to determine if it is rotating; to measure how fast it is moving through space, whether it has a strong magnetic field, and if it is losing material into interstellar space. All this fundamental information, encoded within escaping starlight, is revealed by spectroscopy.

The photosphere is the Sun's final veneer. Just around 100 kilometers deep, the photosphere accounts for a about 0.01% of the Sun's radius. Indeed, if the photosphere was the thickness of a single page in this book, then the Sun, as a sphere to scale, would be a ball about 1.4 meters across. This region, which constitutes the merest fraction of the Sun's total mass, is where the Sun's identity is formed. It is through the interactions that take place within the photosphere, between the outward flowing radiation, atmospheric atoms and the atom's constituent electrons, that a fundamental understanding of stellar properties and star motions can be established. Specifically, it is within the photosphere that absorption lines are produced, and these lines provide the unique fingerprint information on the various kinds of atoms and ions located within a star's outermost region. The top part of Figure 1.2 shows a selection of four stellar spectra.

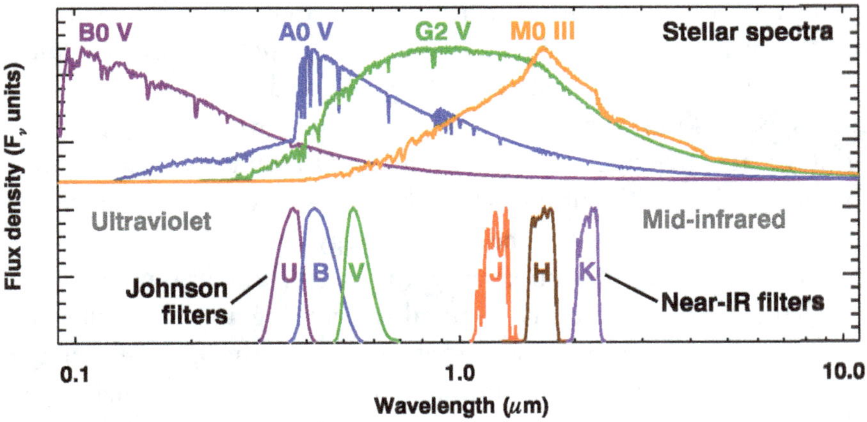

Figure 1.2. (top) Stellar spectra relating to various temperature stars, the B0 V star being the hottest, and the M0 III star the coolest. The Sun is a G2 V star. The diagram shows the run of the measured flux density, that is the energy per meter squared per second per micrometer (μm), against wavelength. (bottom) the wavelength location and transmittance characteristics of various observational filters used by astronomers to gauge and characterise stellar properties (see text for details). Image from: https://users.physics.unc.edu/~gcsloan/research/spectra.html.

The diagram is a plot of the flux density measured at the telescope in the wavelength region from 0.1 to 10.0 μm. Each spectrum shown in the figure reveals characteristic features, these being a peak energy flux in some specific wavelength region, and a series of absorption lines where, at specific wavelength locations, the energy flux noticeably drops (these are the absorption lines formed in a star's photosphere). The reason that the peak flux wavelength position varies is related to the temperature of the photosphere, and will be described shortly. The absorption lines are the result of the quantum mechanical nature of atoms, and specifically how bound electrons interact with the surrounding (photosphere) radiation field.

Atoms are simple things — at least to the mind's eye. In physical reality, of course, they are highly complicated structures composed of quarks, arranged in triplets to give protons, neutrons, and a cloud of electrons. The simplest atom is that of hydrogen, the most abundant element in the Universe, and it is composed of a single proton and a single electron. It is how the electron interacts with its environment and the radiation field in the photosphere that determines the wavelength of the specific absorption lines that might be seen. The strict wavelength arrangement of absorption lines is entirely related to the energy levels associated with the allowed electron orbitals for each specific kind of atom. An electron can move between allowed orbital states by either absorbing energy by interacting with a photon or colliding with another atom, or by losing energy by ejecting a photon that carries away the exact energy difference between the initial and final states. Importantly, the rules of quantum mechanics say that electron transitions are an all-or-nothing affair — either the energy difference for a specific transition is exactly right, allowing a transition to occur, or it is not, and the electron stays where it is. There is, additionally, the underlying quantum mechanical directive that electrons always try to attain the lowest possible energy state allowed. In this sense, if an electron absorbs the exact amount of energy needed to enable a higher energy state, it will go to that higher energy (excited) state, but it will not stay in that exited state forever. Rather, the electron 'seeks' to obtain a lower energy state by emitting a photon that carries away excess energy. Like an elevator in a tall building, the electron travels up and down between fixed floors, but its overarching goal is to remain in the basement.

A look at Figure 1.2 indicates that stars show absorption lines in their spectrum, and these can be recognized and associated with a particular atom based on the results of laboratory studies. In terms of the spectral classification of stars, one of the most important atoms to understand, with respect to absorption line characteristics, is the hydrogen atom. In this case, the wavelength at which an absorption line will appear, when the electron transitions involve the mth and n^{th} states is

$$\frac{1}{\lambda} = R_H \left(\frac{1}{m^2} - \frac{1}{n^2} \right) \tag{1.1}$$

where $m < n$, λ is the wavelength of the absorption line and R_H is the Rydberg constant. For a fixed starting orbital, m, a whole series of absorption lines are produced according to $n = m + 1$, $n = m + 2$, $m = n + 3$, and so on. When $m = 1$, the so-called Lyman series of lines are produced. Named after American physicist Theodore Lyman, these absorption lines fall in the ultra-violet region of the electromagnetic spectrum. When $m = 2$, the so-called Balmer lines are formed. These lines, named after Swiss mathematician Johann Balmer, fall in the visible part of the electromagnetic spectrum. When $m = 3$, the Paschen line series is produced. These absorption lines, named after German physicist Friedrich Paschen, fall in the infrared part of the electromagnetic spectrum. Historically, the hydrogen Balmer lines, along with additional absorption lines from other atoms that also fall in the visible part of the electromagnetic spectrum, have been used to construct star classification schemes. The various schemes have varied greatly over the years, but they are now generally constructed around the specific presence, and so-called strengths, of the absorption lines observed in the visible part of the electromagnetic spectrum. In modern times, spectral observations have been extended over the entire electromagnetic spectrum. With reference to Figure 1.2, however, the A0 stars show the strongest (that is, most distinctive) Balmer lines. B spectral type stars, in contrast, show relatively weak Balmer lines, but demonstrate strong lines due to helium and ionized helium atoms. M spectral type stars show no Balmer lines, but are distinguished based on absorption lines relating to neutral metals (especially titanium oxide). Table 1.1 shows the basic outline of the Harvard spectral classification scheme. The scheme is basically a temperature sequence (see below), with O spectral type stars being the hottest, and the M spectral type stars being the coolest. The refined classification scheme allows for each spectral type to be divided into 10 subclasses, e.g., A0 to A9. For example, the Sun is a G2 spectral type star. Extensions to the basic scheme also allows for the classification of brown dwarfs (through L, T and Y spectral type labels), and for those rare stars showing emission lines, such as the Wolf–Rayet (WR) stars. White dwarfs, as such, do not enter into the Harvard classification scheme, but have their own specific set of diagnostic spectral features.

In addition to using absorption line characteristics (literally the presence of a specific absorption line and its strength) to assign a spectral type, they can also be used to assign a luminosity class. This scale designates a Roman numeral between I to V (with subtypes) to

Table 1.1. The basics of the Harvard spectral classification scheme. The first column indicates the spectral type; the second column indicates the associated Balmer line characteristics; the third column indicates the important classification absorption lines; the fourth column indicates the associated temperature range for each spectral type (see text for details).

Sp. Type	Balmer line characteristics	Key atoms and ions used to distinguish spectral type and subclass	Temperature range (K)
O	weak	He I and He II, ionized Si, O, N, C lines	> 30,000
B	medium	He I, ionized Si and Mg	10,000 – 30,000
A	strongest	Balmer lines, ionized Fe, Mg, Si, Ca	7,500 – 10,000
F	medium	*H* and *K* lines of Ca II, neutral Fe, Cr	6,000 – 7,5000
G	weak	*H* and *K* lines of Ca II, neutral Fe, CN molecule	5,200 – 6,000
K	very weak	Neutral metals Mn, Fe, Si, TiO molecule	3,700 – 5,200
M	none	TiO, CO molecules	2,400 – 3,700

a star, and it is a measure of the so-called collision-broadening that depends upon the characteristic density of a star's outer atmosphere. Stars with low density, extended atmospheres have 'sharper' absorption lines (at a given spectral type), while those with high density, compact outer atmospheres have absorption lines that are 'broader'. The Sun, for example, has relatively broad absorption lines and is designated a luminosity class V star. The largest, most luminous stars have a luminosity class designation of I, whereas ordinary giant stars have a luminosity class of III.

The spectral type and luminosity class designation provides a two-parameter classification scheme based purely on the observed and measurable characteristics of a star's spectrum. It will be seen below that this two-parameter scheme effectively organises stars according to their temperature (via their spectral type), along with their luminosity and physical size (via their luminosity class). These same properties also explain the Hertzsprung-Russell (HR) diagram, which we shall discuss more fully below.

Not only does the spectrum of a star tell us about its temperature, luminosity and size, it also provides information about its compositional make-up. This makes sense, since the absorption lines are due to specific (laboratory measured and understood) atomic transitions. By conducting a detailed analysis to the observed line strengths, a measure of the relative abundance of the various atoms in a stellar atmosphere can be made. Stars, it transpires, are almost entirely composed of hydrogen, with helium coming in as a distant second building block. The Sun's basic composition is found to be 73.81 percent hydrogen, 24.85 percent helium and 1.34 percent all other elements. The so-called metallicity[8] of a star is based upon the logarithm of a star's iron abundance, which (as we shall see later) is taken to increase as the universe ages and subsequent generations of stars form and evolve to their end stage. In this manner so-called populations of stars are defined, with Population I stars being the most recently formed, and having, thereby, relatively high metal content. Population II stars are older, that is, they formed long ago

[8] According to astronomers any element heavier than helium is a 'metal'.

when the galaxy was younger, and they are accordingly distinguished as being metal-poor. A third class, Population III stars, has also been hypothesised, with these being the very first formed stars[9]. Since the composition of the initial universe was purely hydrogen and helium, Population III stars should be distinguished by having zero metallicity. Once again, we shall have more to say about stellar populations later — the point here being that the metallicity (which is used to assign Population membership) is a quantity that is determined from the spectrum of a star. The Sun, for example, is a Population I star. As will be seen later, the Population designation given to a star tells us something about not only its composition (via its iron abundance), but also its formation epoch and its orbital dynamics. Indeed, it transpires that Population I stars are predominantly found within the spiral arms and disc of our galaxy (Figure 1.3), while

Figure 1.3. Our Milky Way galaxy is a so-called barred-spiral galaxy. Population I stars are located within the disc of the galaxy, while Population II stars are found in the thick disc, halo, bulge, and within globular clusters. The Sun (lefthand image) is located within the galactic disc at about 8000 parsecs (246.8 Exameters) away from the galactic center. Image courtesy of ESA.

[9] As of writing these pages, Population III stars remain elusive and no clear detections have been made. The identification of these 'first' stars is, however, one of the major projects being undertaken with the James Webb Space Telescope.

Population II stars are predominantly located within the extended regions above and below the galactic disc, the galactic halo, and the central bulge. Globular clusters (to be described later) are exclusively composed of Population II stars.

In a very broad-brush form, and with respect to Figure 1.3, the Milky Way galaxy is a barred spiral galaxy composed of some 300 billion stars. In terms of structure, it supports a central, spherical core of stars (the bulge), with a radius of about 2 kpc[10], that is composed of a mixture of Population I and II stars. The disk of the galaxy is divided into zones of various heights above (and below) the plane of the galaxy. The spiral arms and regions of new Population I star formation are located within the thin disc (confined to regions within 400 pc of the galactic plane), and this merges into the thick disc (confined to regions within 2 kpc of the galactic plane). The thick disc is predominantly composed of low mass, low metallicity Population II stars. Surrounding the galactic bulge and the disk, out to distances of order 30 kpc, is a spherical halo of old, low mass, Population II stars and numerous globular clusters. At the very center of the galactic bulge sits a supermassive black hole (designated as Sagittarius A*), while the solar system is located within the thin disc at about 8 kpc from the galactic center.

Not only does the spectrum of a star tell us about its chemical make-up, the absorption lines that it displays can additionally be used to determine how fast it is moving with respect to our line of sight. Such information can be used to determine not only how an individual star is moving through space, but, when combined with additional observations[11], how the galactic disk itself is rotating. The line of sight (or radial) velocity of a particular star is gauged according to the Doppler Effect. Described by Christian Doppler in 1842, this effect relates to a systematic shift in the wavelength location at which a specific absorption line is observed. In the laboratory, where all is at relative rest, for example, the first Balmer line of hydrogen

[10] In terms of the Earth's orbital radius about the Sun being 1 astronomical unit (AU), 1 parsec (pc) is equivalent to a distance of 206,265 AU.

[11] These are typically observations of interstellar clouds made with a radio telescope.

(often called the hydrogen α line) falls at a wavelength of $\lambda = 0.656$ microns (see the A0V star spectral in Figure 1.2). If, however, a particular star is moving towards us, then, due to the Doppler Effect, this line will be observed at a wavelength λ_{obs} with $\lambda_{obs} < \lambda$. Conversely, if the star is moving away from us, then the observed wavelength of the line feature will be at some wavelength $\lambda_{obs} > \lambda$. The key point, however, is that the difference between the laboratory-measured and the observed wavelength is directly related to the radial velocity V_R according to the formula

$$\frac{V_R}{c} = \frac{(\lambda_{obs} - \lambda)}{\lambda} \tag{1.2}$$

where c is the speed of light. The convention that goes with equation (1.2) is that a positive value for V_R indicates motion away, while a negative value for V_R indicates motion towards. The common terminology that goes with this convention is that motion away indicates a red shift (with all absorption line wavelengths being shifted to values larger than the measured-laboratory standard at rest), while motion towards is called a blue shift (with all absorption line wavelengths being shifted to values smaller than the measured laboratory standard at rest). The radial velocity, $V_{R,}$ is not actually the velocity with which a star is moving through space (unless it happens to be moving directly away or directly towards us). To determine the space velocity, $V_{S,}$ we need to measure the distance to the star (through its parallax) and determine its so-called proper motion (see below). With this latter information, measurable given enough time and technology, the orbit of a star about the galactic center and its actual speed through space can be determined. While the motion towards or away from us changes the observed wavelength of each absorption line, it is also the case that the speed with which a star is spinning, $V_{spin,}$ is encoded within starlight. In this case, the star's spin velocity is deduced by looking more closely at the profile of a given absorption line — rotation tends to broaden out the profile.

At this stage it seems useful to step back, pause just a little, and reflect upon how much information astronomers can extract from starlight. Indeed, it is truly staggering, and truly marvelous how much information starlight contains. By measuring the energy flux from a star as a function of wavelength (that is by producing its spectrum), we can observe that stars are made of the same elements found on and familiar to us on Earth, and studied in the laboratory — stars are not made of distinctly different materials. We can further use the observational information to set-up a spectral classification scheme (spectral type and luminosity class) that tells us that stars come in a range of temperatures and luminosities. A detailed analysis of the absorption line profiles further enables us to determine the compositional make-up of the stars, finding that they are predominantly composed of hydrogen and helium, but that there is a distinct variation in their metallicity. This latter measure enables us to determine and distinguish between star formation histories and delineate features of galactic structure. Furthermore, we can also use the wavelength at which a specific absorption line is observed to determine if a star is moving towards or away from us, and the detailed study of absorption line profiles can enable the determination of how fast or how slow a star is spinning. While not followed in any detail here, the study of absorption line profiles also enables astronomers to determine if a star has a surface magnetic field (through the so-called Zeeman effect) and whether it is losing mass into space. Starlight is truly packed full of information — provided, that is, one knows how to read it — and while it might be seen as the sparkling nighttime glow that calls the poet to their muse, to the astronomer, starlight is a rich and truly magnificent source of physical data.

As a final entry in this section, look one more time at Figure 1.2. The lower panel is now our focus. The six curves shown in this panel correspond to the wavelength acceptance regions of specific filters that astronomers have adopted to measure starlight by. To the left are the three optical filter profiles for the Johnson UBV system, while to the right are the infrared wavelength acceptance regions for the J, H, and K filter system. By using standardized filter systems astronomers measure the energy flux over a restricted region of the electromagnetic spectrum, and it transpires that these filter-based observations can be calibrated to reveal the luminosity, radius and temperature of a star — this is something that we shall come back to later.

1.3. Star systems

Stars, just like people, can be single, or they can be coupled together in a close relationship. The Sun is a single star, and ever since it first formed, it has been a loner. For all that, the Sun formed in a natal cluster along with many hundreds of siblings. These close cousins are now spread around the galaxy, their paths having evolved and diverged away from the Sun's over the last 4.5 billion years. For the Sun, we have $N(\text{Sun}) = 1$. For the most prominent star in our night sky (at the present epoch), we have, in spite of appearances to the naked eye, $N(\text{Sirius}) = 2$. The bright star we call Sirius is actually a binary system. The brightest star in the system, Sirius A, is an A spectral type star, while its binary companion Sirius B is a diminutive, low-luminosity white dwarf. While Sirius A has been visible to terrestrial observers for many thousands of years, Sirius B was first observed in 1862, although its existence was predicted, on the basis of proper motion measurements (see below), in 1844. In the southern hemisphere, the star α Centauri is a triple system. Again, while α Centauri looks like a single star to the naked eye, it transpires that $N(\alpha\text{Centauri}) = 3$. Indeed, the system is effectively two binary systems. The bright star α Centauri is composed of two relatively close stars, αCen A and αCen B, separated by a distance that varies between 8.4 and 26.6 AU, and this system forms a binary pairing with Proxima Centauri, at a separation distance of some 13,000 AU. There is still some debate as to whether Proxima actually forms a binary with αCen AB, but this is another story not followed in these pages [1]. That αCentauri is a binary star was first realised in 1689; the discovery of its companion, Proxima Centauri, was only made in 1915. Moving back into the northern night sky, the bright star Castor in the constellation of Gemini is a sextuple system, making $N(\text{Castor}) = 6$. In this latter case we have a series of compact binary–binary systems and a slightly more distant binary. In those cases where star systems have $N \geq 2$, the component stars will have an associated set of parameters relating to their orbital properties. In a binary system, for example, the system dynamics will provide data (see below) on the orbital period of the stars, the orbital eccentricity of the stars, and system orientation data.

While the Sun has followed the life of a lonely bachelor, this does not mean that it has led a totally solitary life. There are literally hundreds of billions of stars within the Milky Way galaxy, and close encounters (see below and Chapter 3) between the Sun and other stars have occurred in the past and will continue to occur in the future. At the present epoch, the closest star to the Sun is Proxima Centauri, and it takes a light-ray about 4 ¼ years to travel the distance between the two of them. These two stars are merely passing ships in the night, but at least, for a brief celestially interval, they sail, far apart, but together. This is not always the case, however, and there are times (past and future — see below) when the Sun will undergo a close encounter that is more akin to suffering a pirate raiding party than experiencing a happy encounter involving a courteous dipping of flags.

Risking the wrath, once again, of Walt Whitman's ghost, in terms of the statistics and numbers arrayed, the various survey counts reveal that the fraction of stars that are in binary systems varies significantly according to spectral type. It is generally taken that about 50% of all stars have at least one companion (that is, $N \geq 2$). Something like 80% of O spectral type stars, however, have at least one companion, and some 25% of such stars have two companions (i.e., triple, $N = 3$, systems). At least 15% of F and G spectral type stars are thought to reside in triple systems. Where these companion statistics come into play, with respect to the formation of blue stragglers, will be revealed in Chapter 3.

1.4. Stellar distance measurements

Come Watson. The answer lies in trigonometry
(*The Musgrave Ritual* — Granada TV series)

In order to have any idea how luminous, how large, and how massive a particular star is, an astronomer needs to know how far away the star is, in units directly convertible to meters. Without this measure, all is relative or just speculation. In order to find the distance to a star, the astronomer must first obtain a precise measure of the

Earth's orbital radius[12] about the Sun. This distance, appropriately called the astronomical unit (AU), is the fundamental measure of astronomical distance, and it sets the scale for all subsequent distance measures. Fortunately, in the modern era, the astronomical unit can be measured to an accuracy of just a few meters, and it is determined to be 1 AU = 149,597,870,700 m ± 3 m. With this knowledge of the radius of the Earth's orbit about the Sun, a common baseline for all astronomy measurements can be established: the diameter of the Earth's orbit. Conveniently, the time interval for the Earth to move from one extreme of its baseline to the other is easily gauged as being six months. Being able to determine when the Earth is at the two extreme locations of its baseline enables astronomers to deduce a parallax shift by measuring the relative change in a star's position on the sky. The point about parallax motion, observed when literally looking at the same region of sky from two different viewpoints (the extremes across the Earth's orbital diameter), is that nearby objects will move (shift their position) relative to those situated much further away (which will hardly move). In this manner, for example, one can determine the apparent shift in the position of a star on the sky, with respect to the position of (say) a distant galaxy, and use this measure to determine the star's distance from the Sun. The geometry is simply that of a triangle, and if the apparent shift in the position of a star over a six-month time interval is measured to be S degrees[13], then the distance, D, between the star and the Sun is simply, $D = 1\text{ AU} / \tan(S / 2)$. By measuring S, the distance, D, to the star is then known in astronomical units. For example, if the angular shift in the position of a star on the sky, over a time interval of six months, is measured to be 1/1000th of a degree, then the distance to the star is 115 thousand astronomical units (or 17.1 petameters). Remarkably, for such a seemingly small parallax shift, no star is presently this close to the Sun. The first star to have its parallax accurately determined was 61 Cygni (by Friedrich Bessel in

---

[12]Technically, it is the Earth's orbital semi-major axis and eccentricity that need to be determined.

[13]The angle $P = S / 2$ is called the angle of parallax.

1838) with a six-month angular shift of $S = 1.59 \times 10^{-4}$ degrees = 0.572 mas[14]. This gives a distance of D (61 Cygni) = 720,701.6 AU = 107.6 petameters.

Closely linked to the measure of parallax is the measure of the relative motion of a star across the sky with respect to its visual neighbors on the sky. This motion, which accumulates very slowly, is called the star's proper motion. From the perspective of an observer located upon the Sun (the quintessential 'hot seat'), a given (single) star's proper motion will correspond to some linear (straight line) motion across the sky. Observing from Earth, however, which is in orbit about the Sun, the observed proper motion path of star becomes a wavy line across the sky. Only if no parallax is discernable (which happens when the star is very distant from us) will the proper motion path look like a straight line from Earth orbit. Figure 1.4 shows this combined effect of parallax and proper motion in the Hipparcos satellite observations for the star 70 Virginis. The 'wavelength' of the proper motion path corresponds to the angular shift in the star, due to its physical motion through space, as seen

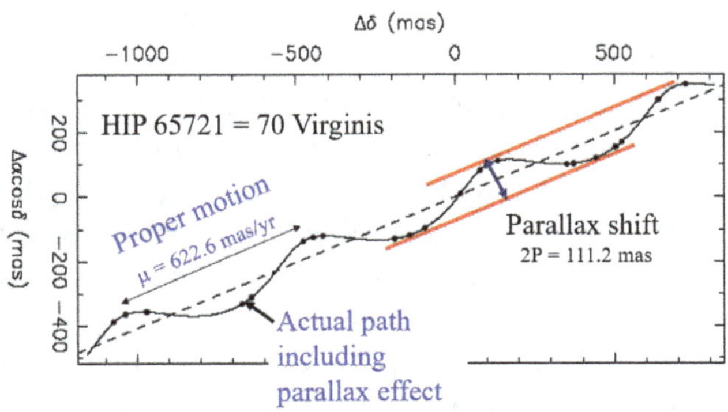

Figure 1.4. The sky motion path (thin curvy line) of the star 70 Virginis as determined from Hipparcos satellite observations. From such information the proper motion μ, and the angle of parallax P can be determined. The scale on each axis is in milliarcseconds (mas). Image courtesy of ESA.

[14]Here we introduce the units of the arc second, with 1 arc second = 1/3600 of a degree. The unit mas corresponds to one thousandths of an arc second.

from the Earth over the course of one-year, while the amplitude of the wave corresponds to the shift in the position of the star, over a 6 month time interval, due to the Earth's movement between the two extreme positions across its orbital baseline. In the case of the Hipparcos satellite observations of 70 Virginis, the proper motion is found to be $\mu = 622.6$ mas/yr, while the angle of parallax is $P = 55.6$ mas (indicating a distance to the star of 3.7 million AU = 0.55 Exameters).

Before moving on to a discussion of the cosmic distance ladder, there is one historical topic, related to parallax measurements that should be tidied away. Specifically, this concerns the definition of the parsec unit (pc). This unit of measure is purely a convenience, and has its origins in a mathematical simplification. The mathematician will look at the equation for D and realize that for very small angles the $\tan(S/2)$ term can be greatly simplified. To the modern reader this is hardly an issue, where all such numbers are calculated on a computer or hand-held calculator. In the early 1800s, however, all calculations had to be made by hand, with pen and paper, and this was a highly time-consuming and frustratingly error-prone process. The mathematical trick that saved a lot of calculating time, however, begins by noting the geometrical identity that $\tan A = \sin A / \cos A$. Now, when A is a very small angle, so cos A approaches a value of 1, and sin A approaches a value of 0. The first limit simply means we can ignore the cos A term once A is very small, since its value is effectively 1. Hence, for small A, $\tan A \approx \sin A$. Now, there is another mathematical trick that enables the sin A term to be expressed by a (technically infinite) power series. Accordingly, $\sin A \approx A - A^3/3! + A^5/5! - A^7/7! +$ higher order terms. This latter step works provided that A is expressed in units of radians. The required unit change is a straightforward conversion step, since by definition 2π radians ≡ 360 degrees. Now, looking at the series expansion term for sin A, it is seen that as A becomes very small, the main contributing term is the first term A. The other terms effectively vanish since one is taking a small number and then making it even smaller by raising it to some power. So, after all this mathematical simplification we end up with the small angle relation: $\tan(A) \approx \sin(A) \approx A$, where A is expressed in radians. The original formula for the distance D can now be simplified as follows: D = 1 AU / tan(P) = 1 AU / [(P / 3600) (2π / 360)] = 20,6264.8 AU / P(arcsec). In the last step the measured angle of parallax is first expressed in arc seconds (which are appropriate units for the angle being measured) and then the resultant angle is converted to radians. Now, rather than having to calculate tan(P) by hand, all the astronomer needs to do is take the deduced parallax angle P (measured in arc seconds) and divide the result into 20,6264.8. Having made a string of simplifications in the mathematics, it now makes sense to make a simplification in the final calculation and set 20,6264.8 AU = 1 parsec. The unit and label of 'parsec' was introduced as a shorthand concatenation of parallax (par) and arc second (sec). Accordingly, after all the simplifications and unit changes, the distance to a star in parsecs can be expressed as

$$D(\text{pc}) = 1 / P(\text{arcsec}) \tag{1.3}$$

The distance to a star is now 1 parsec = 20,6264.8 AU if its angle of parallax is determined to be P = 1 second of arc — that is, if its shift on the sky over 6-months is measured to be S = 2 arc seconds. For the star 70 Virginis, P = 55.6 mas (Figure 1.4), and according D(70 Vir.) = 1 / (55.6×10^{-3}) = 17.99 pc. While the mathematical simplification is no longer needed in the modern era (thanks to electronic calculating devices), astronomers have stuck to the distance unit of the parsec.

The other distance unit that astronomers use is that of the light year (ly). As its name indicates the light year is the distance travelled by light in one year, and is accordingly based upon a good experimental knowledge of the speed of light c and the number of seconds in one year. This definition is a little more complicated than one might at first think, but the time interval corresponds to the so-called Julian year = 365.25 days, and with the speed of light being measured in the laboratory as c = 299,792,458 m/s, 1 ly = 9,460,730,472,580,800 m. In terms of our other distance units, 1 ly = 0.306601 pc = 63,241.077 AU.

The further away a given star is from the Sun, the smaller its angular shift S on the sky becomes, and ultimately a measurement limit[15] will be reached, with the measuring device no longer being able to resolve smaller angles of change. The GAIA spacecraft can determine angular shifts on the sky to a limit of about 250 microarcseconds[16], that is to about 69 billionths of a degree, which translates to an ability to determine parallax distances out to about 1.65 billion astronomical units (some 8 kpc) from the Sun. At this distance GAIA is able to probe the motions, distances and properties of the brighter stars located at the core of our Milky Way galaxy.

Once the proper motion and parallax of a star have been determined, a measure of its motion through space can be gauged. With reference to Figure 1.5, the position of a star is indicated at two locations in space, measured one year apart. The star has a space velocity V_S along a path that makes an angle φ to the line of sight at the time of the first positional measurement (labeled 1). At the first location the star is taken to be a known distance D away from the Sun (i.e., based upon a parallax determination), and its radial velocity V_R is measured (as per Equation (1.3) and the Doppler Effect). At the second location (labeled 2), one year later, the star will have travelled a distance τV_S along its path through space (where τ = 3.15 million seconds is the number of seconds in one year). A tangential velocity

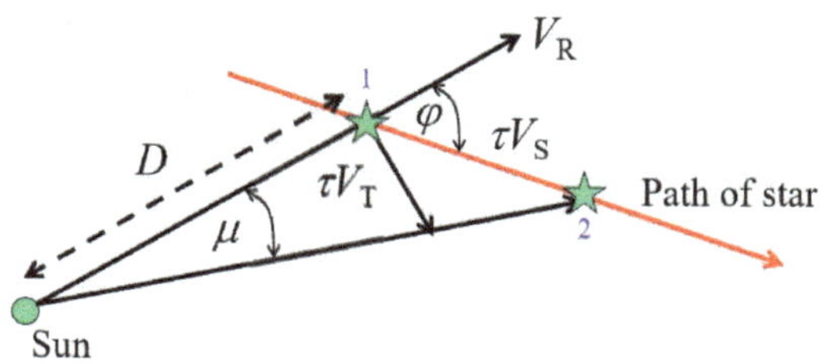

Figure 1.5. The geometry linking the space velocity of a star V_S, to the tangential velocity V_T, the radial velocity V_R, and the distance $D = 1 / P$, where P is the parallax in arcseconds, and the proper motion μ.

[15] The limit is set not just by the angular measurement limit, but also by how bright the star is. The further a star is away from the Sun, the fainter it will appear, and accordingly it can drop below a system's threshold for detection.

[16] This is a little arbitrary, but assumes angular shifts that are at least 10 times larger than the system resolution limit of 25 micro arc seconds.

V_T has also been introduced, with the idea that this is the velocity component of the star at right angles to the line of sight at the time of the first observation. We can now establish a number of trigonometric relationships. Specifically, tan (μ) = τ V_T / D, tan (φ) = V_T / V_R, and from Pythagoras we have $V_S = \sqrt{V_R^2 + V_T^2}$. Using the same small angle approximations for the tangent equations as applied in the parallax–distance relationship, inserting τ and applying appropriate unit conversions, we find

$$V_T = 4.74 \frac{\mu}{P} \equiv 4.74 D(\text{parsec}) \mu(\text{arcseconds per year}) \tag{1.4}$$

where P is the angle of parallax, as measured when the stars is at point 1 [with D(parsecs) = 1 / P(arcseconds)]. The odd looking constant in Equation (1.4) comes about because the tangential velocity is expressed in units of km/s.

In summary, so far, by measuring the parallax and proper motion of a star, and combining these observations with a measure of the star's radial velocity (from the wavelength shift in the absorption lines present in its spectrum, and Doppler's formula), we obtain knowledge about how far away it is (in units of meters, kilometers, lightyears, or parsecs) from the Sun, along with details of its speed (in km/s) through space, and the direction in which it is moving through space.

While direct, parallax-based measurements, can be used to determine the distance (for at least the brighter stars) out to distances of some 10,000 pc from the Sun, there is invariably a technology limit (literally the limit of angular resolution) that truncates its use to more distant objects. The distances to other stars can nonetheless be gauged by setting-up calibration schemes based upon those stars that do fall within the reach of parallax techniques. In this manner, astronomers have set-up a number of schemes that adopt certain fixed or known properties of specific types of stars (and further out into the universe, galaxies). These schemes are based upon the idea of defining standard candles and standard rulers (Figure 1.6). That is, a calibration scheme is established from those objects that do have parallax distance measures and an object-specific calibrated

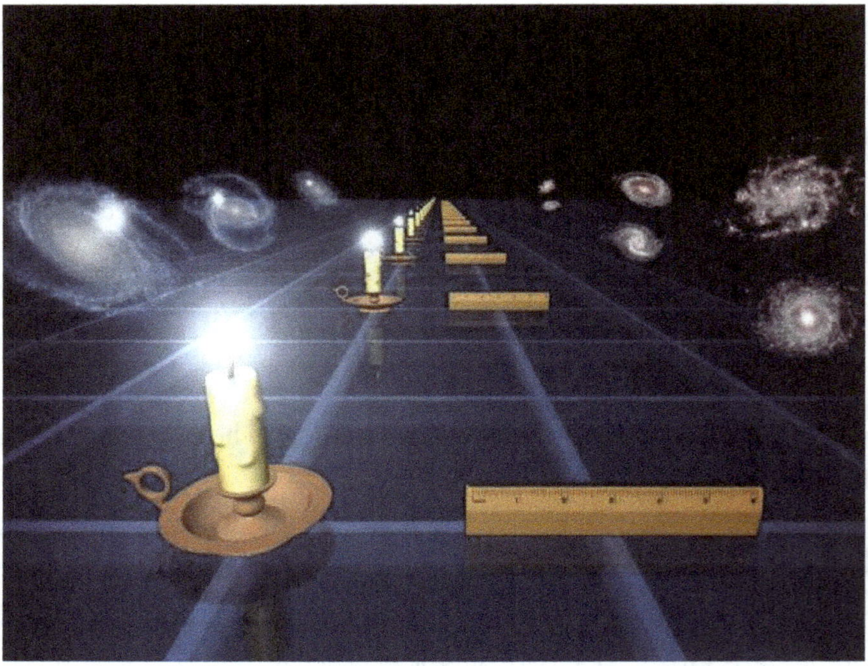

Figure 1.6. The cosmic distance ladder proceeds, to the edge of the observable universe, through the calibration of standard rulers (objects of known size) and the calibration of standard candles (objects of known brightness). In each case, the point is that the further away a given calibrated object is the smaller will its angular size be, and that its brightness must diminishes according to the inverse square law. Image courtesy of NASA.

quantity, either a brightness or a size characteristic, that is taken to hold fixed for all objects similar to those observed in the calibration group. This calibrated knowledge is then carried forwards, or outwards, to stars beyond the direct parallax limit. This method not only requires the calibration of a physical quantity (energy output, spectral type, or size), it also requires a classification scheme for identifying stars of different spectral type. Standard candles and star types will be introduced in Section 1.6, and for the moment the focus will be on standard rulers. The idea of this calibration is that an object of a fixed and known size, d, appears smaller (that is, its angular size is smaller) the further it is away. For all that, however, provided the angular size φ can be measured, and d is taken as a known quantity for the object under study, then distances D can be found. Once again, straightforward trigonometry allows for the determination of distance from the formula $\tan(\varphi / 2) = (d/2) / D$. Since φ is measured and d is known (that is, d has been calibrated), the distance D can be determined even if the parallax is too small to measure. Once again, the small angle approximation can be applied and in the case of a standard ruler distance measure, $D = 206264.8\ d / \varphi$ (arcsec). Once again, the units of distance can be expressed in meters, parsecs, or lightyears, but note that the same distance unit applies to both d and D. The formula relating distance to brightness is a little more complicated, and requires a little more discussion, and this will be presented in Section 1.6 below.

1.5. Blackbody radiators

> *From the standpoint of our investigations, a star is primarily a generator of radiation, and all that we know about it is learned from observations made of this radiation.*
>
> (W. H. McCrea. *Physics of the Sun and Stars*. Hutchinson's University Library, London, 1950)

There is no such thing as a perfect blackbody radiator, but as a theoretical construct, derived according to quantum mechanics, thermodynamics and electromagnetic radiation theory, such imaginary bodies provide useful relationships that enable astronomers to deduce fundamental stellar properties[17]. The basic idea behind a blackbody radiator is that all hot objects radiate energy into space, at all wavelengths of electromagnetic radiation. What the theory allows for is an evaluation of the wavelength λ_{max} at which, for a given temperature T, the greatest energy flux is produced, and for an evaluation of the total amount of energy emitted over all wavelengths per unit area per second, F. The energy flux emitted at a specific wavelength was first fully described by Max Planck in the

[17]The closest *natural* example to a perfect blackbody radiator is that of the cosmic microwave background — but even this is not perfect. Indeed, we would not exist if it was perfect.

closing years of the 19th century, and constituted the first successful application of quantum mechanics. Figure 1.7 shows a set of Planck curves corresponding to three blackbody radiators with temperatures of 18,000, 6,000, and 4,000 Kelvin[18] respectively. Each curve specifies the energy flux at each wavelength parameterized according to the temperature. Indeed, the key point about blackbody radiators and Planck curves is that they are fully described once the temperature T is specified. Planck's formula, which describes the spectral density $B(\lambda, T)$ of a blackbody radiator, is worth seeing (at least once, given its importance), although we do not make any great use of it in following chapters. Here it is:

$$B(\lambda,T)=\frac{2hc^2}{\lambda^5}\frac{1}{e^{hc/\lambda kT}-1} \tag{1.5}$$

The spectral radiance corresponds to the energy that a blackbody radiator emits in the form of electromagnetic radiation per unit area,

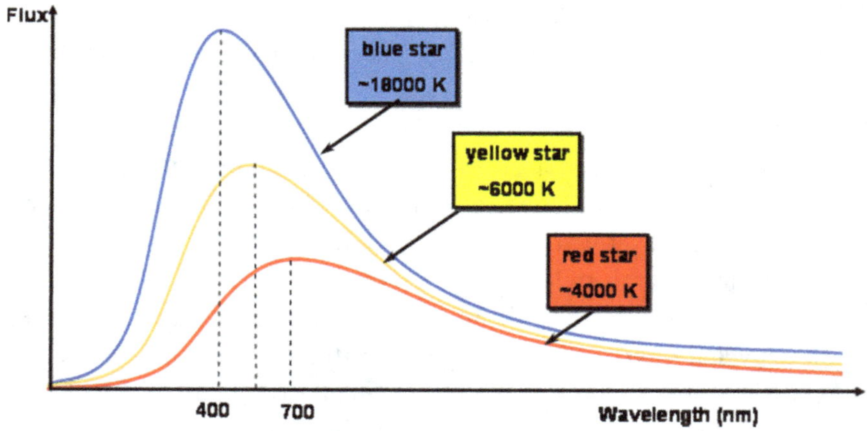

Figure 1.7. Planck curves for blackbody radiators of temperature 18,000, 6,000, and 4,000 K respectively. The vertical dashed lines indicate the wavelength of maximum energy flux. Image courtesy Swinburne University of Technology.

[18] The Kelvin temperature scale takes its starting point as absolute zero, with 0 K being equivalent to –273.15 °C.

per solid angle, at a given wavelength, and is expressed entirely in terms of two fundamental physical constants: the temperature, T, and the wavelength, λ. The formula is historically significant since it introduced the idea of quantized energy[19] into the otherwise unsuspecting world of classical physics. Key to progress, however, is the fact that astronomers have designed and constructed detectors that can measure the energy flux at some specified wavelength. Furthermore, Planck's formula is the basis behind the formulation of Wien's law and the Stefan–Boltzmann law — and it is these laws that we will look at next.

As is evident from Figure 1.7, a blackbody radiator behaves in a very specific manner. The energy flux at very short wavelengths and at very long wavelengths of electromagnetic radiation is small. There is, however, a maximum energy flux at some wavelength λ_{max}, the value of which is determined by the temperature T alone. Indeed, this constitutes what is known as Wien's law, first derived experimentally by physicist Wilhelm Wien, that states that $\lambda_{max}\, T$ = constant. Planck, in his pioneering quantum explanation of blackbody radiators, showed that the constant in Wien's law is expressible in terms of fundamental constants[20] as 2.897772×10^{-3} mK. What Wien's law tells us is that the hotter a blackbody radiator is, the smaller the wavelength attributed to λ_{max}. And conversely, cooler objects have λ_{max} located at longer wavelengths. Taking stars to be approximate blackbody radiators, the hottest stars will have a λ_{max} value towards the blue wavelength end of the visible part of the electromagnetic spectrum (sometimes even in the ultraviolet part). Accordingly, such stars will appear bluish in color since that is the wavelength at which most of the energy flux is emitted. Conversely a cool star will have λ_{max} situated towards the red wavelength, sometimes even the infrared, part of the electromagnetic spectrum. Such

[19] The energy E is expressed in terms of its associated frequency f and Planck's constant h, with $E = h\,f$.

[20] These include Planck's constant, the speed of light, and the Boltzmann constant. Technically speaking, Wien's law follows by setting the differential of equation (1.5), with respect to wavelength, to zero.

stars will appear reddish in color accordingly. The Sun has its λ_{max} at a wavelength of about 500 nm, in the yellow region of the electromagnetic spectrum. If stars were perfect blackbody radiators, then a straightforward determination of their temperatures would follow from the measurement of λ_{max} and an application of Wien's law. That stars are not perfect blackbody radiators is, in fact, a good thing, since we would not be able to determine their compositions (as per Figure 1.2), but it does complicate the measurement of their actual temperatures. This being said, Wien's law and an estimate of λ_{max} will provide a ballpark determination of T. A finessed determination of a star's temperature can only be derived from the detailed study of the absorption line profiles that it displays in its spectrum.

A second important result that stems from the theory of blackbody radiators is that the total energy flux (that is, the energy emitted per second per unit area) over all wavelengths, F, is given by the Stefan–Boltzmann law, with $F = \sigma T^4$, where σ is the Stefan–Boltzmann constant (which again can be expressed in terms of various fundamental constants). The total flux F, over all wavelengths of electromagnetic radiation, is the area under a given Planck curve[21] for some temperature T. Where this becomes useful in the case of astronomy, is that taking a star to be spherical blackbody radiator of radius R allows for an estimate of the luminosity, L, to be made. The luminosity, defined as the total amount of energy radiated into space per second, is accordingly, L = (surface area of star) $\times F = 4 \pi R^2 F$. This provides a convenient relationship between the luminosity, radius and temperature of a star:

$$L = 4 \pi R^2 \sigma T^4 \tag{1.6}$$

This formula is generally more useful in determining the radius, R, of a star, since L and T can be determined by measuring the energy flux through various sets of filters and afterwards, introducing a series of calibration steps. The radius of a star can be determined

[21] That is, the total flux follows by integrating equation (1.5), with respect to the wavelength running from zero to infinity.

directly if it happens to be in an eclipsing binary system (see below), or it is close enough to be resolved into a disc directly[22].

1.6. Fluxes to magnitudes

To the logician all things should be seen exactly as they are
(*The Adventure of the Greek Interpreter*)

Greek philosopher Hipparchus introduced the idea of gauging star brightness according to a magnitude scale in c. 150 B.C. The idea was straightforward, and was simply based upon an imposed arbitrary scale that assigned magnitude 6 to the faintest stars visible to the eye, and magnitude 1 to the brightest stars visible to the eye. The Sun was not included in the scale since it was not clear, at that time, that it had any characteristics in common with the stars in the sky. The brightest star in the sky in Hipparchus's day, and still in ours, is Sirius. The Hipparchus scheme, however, in spite of its simplicity, is problematic — it is the case that any scale based upon an individual observer's judgement is doomed, ultimately, to fail. What one person judges as a magnitude 2 star, another might call magnitude 3, or magnitude 2.7, and so on. For the magnitude scale to be truly useful, it has to be based upon a quantity measurable with a calibrated instrument. This was a topic developed by Norman Pogson in the late 19th century.

The quantity that can be measured and ultimately expressed as a magnitude, m, is the energy flux f of electromagnetic radiation received from a given star, over a specified wavelength region, at a telescope's detector. The energy flux will be measured in units of Joules per square meter per second[23]. These units will allow us, eventually, to figure out a useful relationship between the total energy output from the surface of a star and its distance (see below). For the

[22] Only a very few stars fit into this category — there is the Sun, of course, as well as the supergiant star Betelgeuse, and the North Star (Polaris) is another.

[23] A caveat at this stage is to remember that fluxes are measured over a specific wavelength range, rather than all wavelengths of the electromagnetic spectrum.

moment, however, the magnitude scale introduced by Pogson is based upon the relationship

$$m = -2.5 \log f + C \tag{1.7}$$

where C is a constant to be calibrated. The conversion between a measured flux f and a derived magnitude m is entirely arbitrary in principle, but Pogson wanted to marry his system of magnitude determinations with that of Hipparchus. In this respect, Pogson argued that if two stars differed by a factor 100 in their measured fluxes — that is, $f_1 = 100\ f_2$ — then, by definition, they must differ by 5 magnitudes in brightness. It is a straightforward calculation to verify this numerically, but to make the magnitude system useful one must also define and introduce a set of standard stars. These are stars (selected from across the entire sky) that everyone agrees have some specified and constant magnitude. Pogson suggested, for example, that the star Altair = α Aquilae be designated a magnitude $m = 1$ star. It is by this last step that the magnitude scale, based upon actual flux measurements, can be calibrated. The key point being that anyone, at any observatory, at any time, can measure the flux of Altair[24] and that of some different star, star S*, and use Equation (1.7) to determine the actual magnitude of star S*. And the magnitude deduced for S* will be the same for all observers. In many ways, the magnitude scale is an anachronism. There is technically no specific advantage to talk about star magnitudes, given that the key physical quantity, the quantity that is actually being measured, is the energy flux. For all that, there is a certain numerical convenience to work with magnitudes, and indeed, astronomers have generally stuck with the process[25].

Before looking at the consequences of Pogson's scheme, a few comments on the idea behind standard candles should be made.

[24] There is, of course, the caveat that Altair must be above the horizon at nighttime.

[25] It is interesting to note that only observational astronomers have stuck with the magnitude conversion convention. Radio astronomers, and astronomers using X-ray or gamma ray satellite data, in contrast, simply report measured fluxes.

Indeed, standard candles relate total energy output in the form of electromagnetic radiation to the distance an object is away from us, and to the flux that astronomers actually measure with their instruments. As indicated in Section 1.5, the luminosity L is defined as the total amount of energy radiated into space per second over all wavelengths of light. The energy (assumed to be emitted in an isotropic manner) radiated in, say, one second will move into space (at the speed of light), distributed over the surface of sphere. When the expanding front intercepts the Earth, the sphere has radius D, where D is the distance to the star. Furthermore, the flux f that the astronomer will measure, at the distance D, will be the total amount of energy radiated into space per second, the luminosity L, spread over the entire surface area $4\pi D^2$. Accordingly, $L = 4\pi D^2 f$, and this can be rearranged to give the distance as:

$$D = \sqrt{L/4\pi f} \tag{1.8}$$

In Equation (1.8) the flux f is measured instrumentally through a telescope and detector, and L is calibrated from those known and identified objects that are close enough to have a measurable parallax. The question now is how we perform the calibration steps. Recall, two calibration steps are needed. The first enables the recognition of similar objects — a step, in the case of stars, involving spectroscopy and the determination of spectral type and luminosity class (as described in Section 1.2). The second step requires the calibration of the luminosity appropriate to each star according to spectral type and luminosity class when the parallax can be measured (that is, when the distance D is determined directly). Figure 1.8 indicates the steps in the program that we shall follow. The flow chart begins with an observational step and then proceeds, through levels 3, 4 and 5, to link the observations to equations and results associated with blackbody radiators. This linkage eventually enables the determination of a star's luminosity, temperature, and radius. In this manner observations are used to deduce physical quantities that define the external properties of stars, that can then be calibrated against observations relating to spectral type and luminosity class.

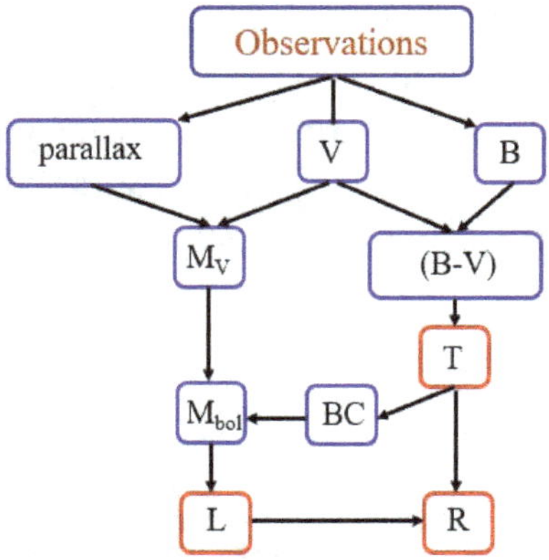

Figure 1.8. The pathways that take the observations of stellar parallax and Johnson V and B flux measurements to a determination of a star's energy output (luminosity *L*), temperature (*T*) and size (*R*). See text for details.

The calibration procedure begins with the observations that work to measure the parallax and the V and B magnitudes. The parallax observations provide a distance to the star being studied, while the magnitudes relate to flux measurements made through two of the Johnson UBV filters — as described earlier with respect to Figure 1.2. In addition to the observations indicated in Figure 1.8, the spectral type and luminosity class of each star being analysed is deduced from their measured spectra.

The V and B terms in Figure 1.2 are based upon flux measurements made through two of the Johnson filters: the V and B filters. This provides us with a measure of a star's apparent visual magnitude V when seen through the V filter, and its apparent B magnitude when seen through the B filter. These magnitudes are called apparent magnitudes since they are based entirely upon the flux that is recorded at the telescope. To take this process further, we need to introduce the idea of magnitude measurements at different distances. Now, this, of course, is not actually possible — astronomers can only measure the flux at one distance: the distance that the star is away from us. However, we can ask what the flux and magnitude of a star would be at some alternative distance from the observer.

Looking again at Equation (1.7), the apparent magnitude V of a star is related to the measured flux, f, and we have seen that the flux can be expressed in terms of a star's distance, D, and luminosity, L, with $f = L / (4\pi D^2)$. A little algebra now allows a derivation of what is called the absolute visual magnitude M_V, based upon the idea that it is evaluated according to the flux that would be measured (by an imaginary observer) a distance of 10 parsecs away from our star of interest. Accordingly, with the distance d expressed in parsecs,

$$M_V = V - 5\log\left(\frac{D}{10}\right) \tag{1.9}$$

In deriving Equation (1.9), it has been implicitly assumed that the measured flux f is the true and unadulterated flux, and that it has not been reduced in any manner between the star and the telescope detector. This is not, of course, generally true. Indeed, we live in the dusty disk of the Milky Way galaxy, and interstellar dust will act to modify the radiation that traverses through it. By chance, the physical size of interstellar dust is comparable to the wavelength of blue light: the dust is about 5×10^{-7} meters in size. As a consequence of this, blue light is strongly scattered (redirected away from the line of sight) by interstellar dust. The more dust there is in the line of sight of the telescope, the greater the amount of blue light that is scattered, and the greater the reduction in the received flux. In effect, starlight is reddened as it moves through dust-filled space. Unfortunately for astronomers, however, the dust in the disk of the galaxy is not uniformly distributed; some regions have high extinction, others very little.

The effect of interstellar reddening is generally accounted for by an extinction term, A, expressed in magnitudes per kiloparsec. That is, the magnitude increase is related to the distance of the star being observed and the column-averaged amount of dust between the star and the telescope. A value of $A = 0$ indicates that there is no dust in the line of sight, and accordingly, no magnitude correction is needed. A value of $A = 0.1$, indicates that for every 1,000 pc, in the direction along the line of sight, the magnitude increases (becomes more positive) by 0.1 magnitudes. This extinction term modifies Equation (1.9) to read,

$$M_V = V - 5\log(D) - DA - 10 \tag{1.9A}$$

Note in Equation (1.9A), the distance is expressed in kiloparsecs (this accounts for the –10 constant term). There is no simple procedure or equation that describes the extinction term *A*, and astronomers have to rely on various star-counting methods (literally counting the number of stars seen in a given area on the sky) in the line of sight to estimate its value and variation in a particular direction within the galaxy.

At this stage it hardly seems that we have made any headway, but once again, the idea is calibration. The observations of the flux and parallax allow us to determine M_V, and the additional spectrum observations — the spectral type and luminosity class — allow us to associate a specific absolute magnitude with the observed spectral characteristics. Sun-like stars, for example, are classified as G2 V (spectral type and luminosity class[26]) stars, and now one thing that we can say about all Sun-like stars is that they have absolute magnitudes M_V(G2 V stars) = + 4.83. This seemingly neutral step of introducing the absolute visual magnitude will eventually allow for the determination of a star's luminosity.

To determine the luminosity of a star, we first need to think a little more about the flux measurements that are being made. The flux measured at the telescope *f* through the V filter allows for the determination of the apparent visual magnitude *V*, but the V filter only allows radiation from a restricted part of the electromagnetic spectrum through to the detector (recall Figure 1.2). Recall that the luminosity, by definition, is evaluated from the total flux *F* radiated at all wavelengths of the electromagnetic spectrum (see Figure 1.9).

---

[26] Note that I am using V here in two ways. The first corresponds to a visual, V magnitude, while the second is the Roman numeral 5 for luminosity class. It should be clear where each usage applies.

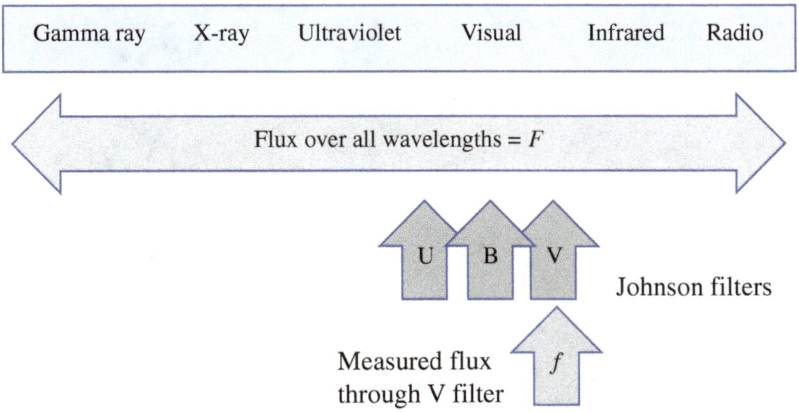

Figure 1.9. The total flux F is that measured over the entire wavelength range of the electromagnetic spectrum ($0 < \lambda < \infty$), from the long wavelength radio waves to the very short wavelength gamma rays. The flux actually measured at the telescope typically corresponds to wavelengths between the ultraviolet and infrared part of the electromagnetic spectrum.

Accordingly, the flux measurements that are actually made provide a lower bound for the total flux: $f < F$.

From the total flux F, measured over all wavelengths, we can construct an apparent bolometric magnitude $m_{bol} = -2.5 \log F + C$, and with this we can construct an absolute bolometric magnitude M_{bol}. From the absolute bolometric magnitude, we can proceed to the direct determination of the luminosity L with

$$M_{bol} = 4.79 - 2.5 \,\mathrm{Log}\,(L/L_{\odot}) \tag{1.10}$$

where the constant term of 4.79 corresponds to the absolute bolometric magnitude of the Sun, and where $L_{\odot}$ is the Sun's luminosity. Since the bolometric flux is not something that can be measured easily, astronomers have introduced the trick of linking the absolute visual magnitude M_V, as derived from flux measurements made with an optical telescope, to the absolute bolometric magnitude M_{bol} through the so-called bolometric correction, $BC = M_V - M_{bol}$. The bolometric correction term is a quantity that varies according to the temperature of a star, and is derived according to detailed computer

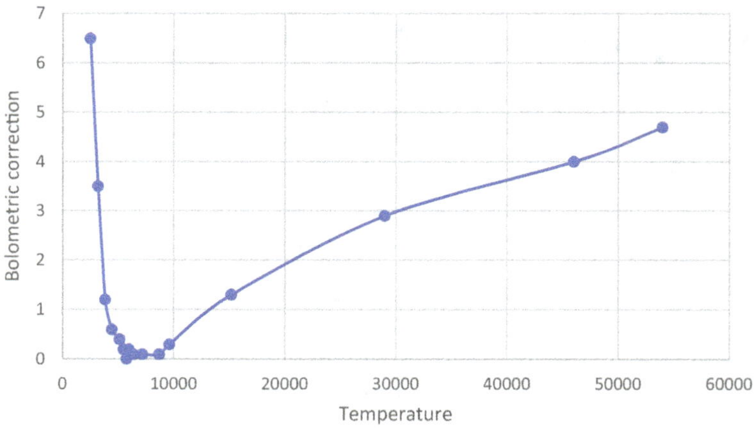

Figure 1.10. The variation of the bolometric correction (BC) with respect to temperature (T). Data points are from Table 1.2.

models of stellar atmospheres (Figure 1.10). Once again, the bolometric correction can be calibrated against spectral type and luminosity class.

The reason that the bolometric correction is large at both low and high temperatures is explained by Wien's law for blackbody radiators (see Figure 1.7). Low temperature stars emit most of their radiation at infrared wavelengths, while very hot stars emit mostly in the ultraviolet spectrum. Accordingly, since the flux is being measured through a filter that selects out radiation from the visible light part of the electromagnetic spectrum, the biggest corrections will be needed at the highest and coolest temperatures. Stars that emit most of their radiation in the visible part of the electromagnetic spectrum (surface temperatures of order 5,500 K) will require the least amount of correction for emission at longer and shorter wavelengths of electromagnetic radiation, with the BC term being of order zero (note, however, BC is never exactly zero, since $f < F$). Various formulae have been developed to describe the variation of the bolometric correction in terms of temperature T, and a commonly used (semi-accurate) expression gives

$$BC = -42.5 + 10\log(T) + \frac{2900}{T} \tag{1.11}$$

In order to determine the luminosity (from the parallax, M_V and BC terms — see Figure 1.8) we now need a means of finding the temperature of the outer layers of a star.

In principle, the temperature of a star can be estimated from its spectrum through Wien's law by treating the star as a blackbody radiator and finding the wavelength of maximum energy flux[27]. Another approach is to measure the flux through two different filters, for example, the Johnson B and V filters, and link these to the fluxes expected from a blackbody radiator of temperature T. In principle, the difference between the fluxes measured over two different wavelength regions can be uniquely matched to a specific Planck curve (recall Figure 1.7), and this provides a specific temperature, the so-called effective temperature, of a star. The temperature of a star is accordingly derived from a calibrated formula between the temperature and the so-called color index (B – V). In this manner, the flux is measured through the B and V filter, and the B and V magnitudes are then computed. A straightforward relationship between T and (B – V) is then given by the relation

$$T = \frac{6957}{(B - V) + 0.7} \tag{1.12}$$

With the temperature now available from calibrated and measured quantities, the pathways indicated in Figure 1.8 can be completed, with observations of parallax (giving distance) and flux measurements through the B and V filter providing for estimates of a star's luminosity (total energy output per unit time), radius[28] and temperature. Table 1.2 provides a summary of the calibrated and basic physical quantities of main sequence (luminosity class V) stars.

The information displayed in Table 1.1 provides us with an overview of the basic, that is fundamental, properties of single stars.

---

[27] A much more accurate method is to model the absorption line characteristics in the spectrum of a star — but this takes us well beyond our current needs.

[28] The radius follows from the substitution of L and R in equation (1.6).

Table 1.2. Characteristics of main sequence (luminosity class V) stars. The first column indicates spectral type. The mass, radius, and luminosity appropriate to each spectral type are given in solar units, while the temperature is given in Kelvin. The mass estimates are derived from the study of binary star systems.

SP type	M ($M_\odot$)	R ($R_\odot$)	L ($L_\odot$)	T (K)	M_V	BC	(B-V)
O5	60	15	800,000	46,000	−5.6	4.0	−0.32
B0	15	6.0	16,000	29,000	−4.0	2.9	−0.30
B5	5.5	4.0	750	15,200	−1.1	1.3	−0.15
A0	2.5	2.5	65	10,000	0.8	0.3	0.0
A5	2.0	2.0	25	8,700	1.8	0.1	0.19
F0	1.5	1.5	10	7,200	2.4	0.1	0.19
F5	1.4	1.2	4.0	6,400	3.3	0.1	0.41
G0	1.1	1.05	1.5	6,000	4.2	0.2	0.59
Sun	1.0	1.0	1.0	5780	4.8	0.07	0.65
G5	0.95	0.98	0.7	5,500	4.9	0.2	0.69
K0	0.8	0.9	0.4	5,150	5.9	0.4	0.84
K5	0.6	0.8	0.2	4,450	7.5	0.6	1.08
M0	0.5	0.6	0.08	3,850	8.9	1.2	1.41
M5	0.3	0.4	0.01	3,200	14.5	3.5	1.80

Table 1.2 arranges the characteristics of stars according to mass, radius, luminosity and surface temperature. It also provides the calibration information needed to find the distance to any main sequence (luminosity class V) star, anywhere in the observable Universe. Provided the spectral type and apparent visual magnitude of a star can be measured, we have information, through the calibrated numbers in Table 1.2, to calculate the distance to that star (via Equation 1.9). Furthermore, we also have some idea of the star's physical properties in terms of its mass and radius. Additionally, as will be seen later, we also have some idea as to the age of the stars that we see, based upon the determination of spectral type, and luminosity class, along with the information about initial mass (second column in Table 1.1).

1.6.1. *Cepheid variables*

One of the most important standard candle calibration steps in determining stellar distances involves the Cepheid and RR Lyrae variables. At some level, all stars vary in brightness, but the Cepheids do so in a very particular and distinctive way. Two types of Cepheid variables are recognized — the classical or Type I Cepheids and the W Virginis or Type II Cepheids — but the key point is that they pulsate, meaning that they undergo steady periodic variations in size and brightness, with the latter being calibrated against their mean luminosity. Figure 1.11 illustrates the luminosity-period relationship for the Cepheids and the RR Lyrae stars. The observations indicate that in the case of the Cepheids, the longer the pulsation period, the greater the luminosity, whereas for the RR Lyrae stars, the luminosity is essentially constant, irrespective of the pulsation period. A characteristic calibration of the period-luminosity relationship for classical cepheids is

$$\text{Log}\,(L/L_{\odot}) = 2.57 + 1.27\,\text{Log}\,(P) \tag{1.13}$$

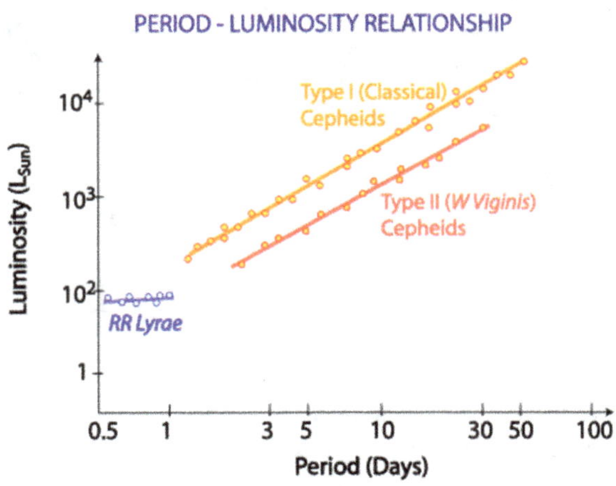

Figure 1.11. Luminosity-Period relationships for Cepheid variables and RR Lyrae stars. Image courtesy of Australia Telescope National Facility.

where the pulsation period P is measured in days. With a pulsation period of 10 days, a classic cepheid variable has a characteristic luminosity 7,000 times greater than that of the Sun.

As can be seen on the vertical axis of Figure 1.11, the Cepheids and RR Lyrae stars are intrinsically luminous, being in the order of 100 to 10,000 times more luminous than the Sun, and accordingly, they can be identified at vast distances across the Universe. Not only are these variables bright, they also vary in brightness in a distinctive and characteristic manner (see Figure 1.12) — this is key to their identification. The brightness variations (the light curve) show a rapid rise to a maximum followed by a more gradual decline to a minimum — this variation is distinctive and different from other variable stars. The light curve is used to confirm the fact that a Cepheid or RR Lyrae star has been found, and the light curve provides a measure of the pulsation period — that is, the time to repeat a pulsation cycle. Once the period has been determined, the corresponding, calibrated luminosity (or absolute visual magnitude M_V) is found, and the standard candle formula then provides a distance measure — as per Equation (1.9) — once the flux has been measured.

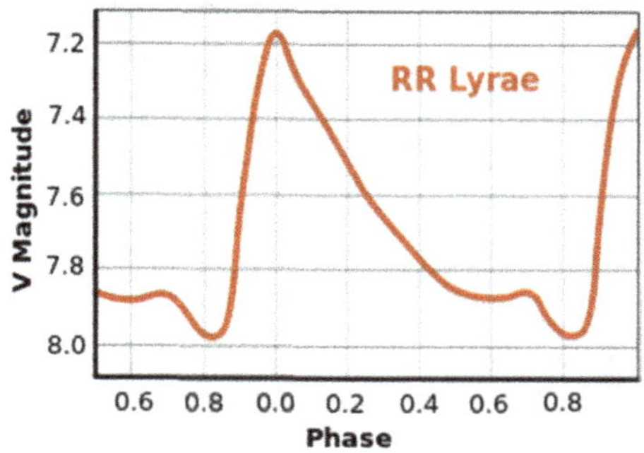

Figure 1.12. The characteristic light curve for an RR Lyrae star. In this figure the apparent visual magnitude is plotted against phase ϕ, where $0 \leq \phi \leq 1$ corresponds to the time in units of the measured pulsation period. The characteristic shape of an RR Lyrae (and Cepheid variables in general) light curve is that of a rapid, short-duration rise to a maximum brightness, followed by a much longer decay time to minimum brightness. Image courtesy of Wikipedia: https://commons.wikimedia.org/wiki/File:Rr_lyrae_ltcrv_en.svg.

Cepheid variables are of fundamental importance in establishing the cosmological distance scale since these stars are bright enough to be seen in distance galaxies. The RR Lyrae stars are also of fundamental importance in determining the distances to globular clusters (see below), and an understanding of these latter structures is important with respect to identifying and understanding blue stragglers.

1.7. From orbits to mass and radius

Astronomers have but one way to directly determine the mass of a star, and that is to observe stars within a binary system. The important point about stars in binary systems is that their masses are encoded within the systems orbital properties, and that these are properties that can be observed and measured. The two stars in a binary system will follow elliptical orbits about a common center of mass. If the distance from the center of mass to star 1 is a_1, and the distance to star 2 is a_2, then the first moment of the system gives the relationship

$$M_1\, a_1 = M_2\, a_2 \tag{1.14}$$

Provided the distance to the system is known (from a parallax measurement or standard candle calibration), the physical distances a_1 and a_2 can (in principle[29]) be measured. Importantly, the stars in a binary system will also satisfy Kepler's 3rd law, which provides an estimate for the sum of the masses, as

$$M_1 + M_2 = \frac{4\pi^2}{G}\frac{a^3}{P^2} \tag{1.15}$$

where $a = a_1 + a_2$, G is the gravitational constant and P is the orbital period. With Equations (1.14) and (1.15), there is enough information to determine each individual mass, that is, once the system

[29] The term 'in principle' is used since the system has to be close enough to the Sun that the angular separation between the two stars can be measured from Earth. No one ever said astronomical measurements are easy.

properties (distance, orbital period, a_1 and a_2) have been measured. Furthermore, once the masses have been determined, they can then be calibrated against the specific spectral type and luminosity class of the stars within the system. Indeed, for luminosity class V stars — that is, main sequence stars — there is a distinct correlation between star mass and star luminosity (Figure 1.13). The luminosity follows a clear power law relationship with mass, with $L \sim M^{\alpha}$ and $1 \leq \alpha < 4$. The $\alpha = 1$ condition applies to the most massive stars, with masses in the order of 50 to 100 times the mass of the Sun (not actually shown in Figure 1.13), while $\alpha = 3.5$ applies to stars within the mass

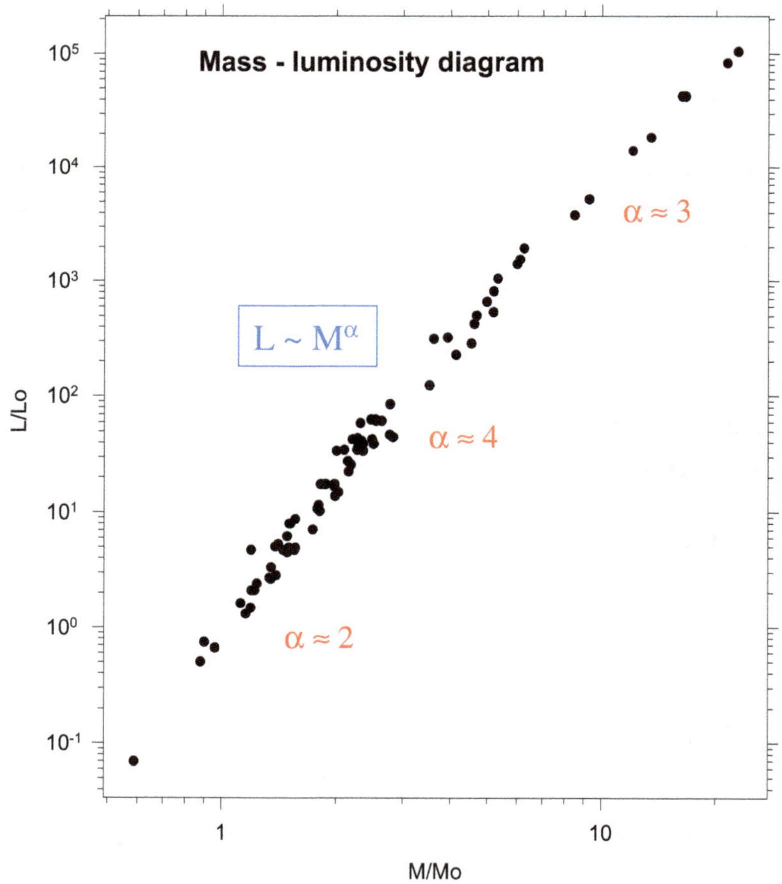

Figure 1.13. The luminosity-mass relationship for main sequence (luminosity class V) stars. Data points are from [2].

range of 0.5 to 5 times that of the Sun. The close correlation between stellar mass and luminosity on the main sequence was a pivotal astronomical discovery, and it proved to be of major importance with respect to the rise of stellar astrophysics in the early 1900s. Indeed, explaining why such a power-law relationship between stellar mass and luminosity must apply was one of the early triumphs of the astrophysical theory of stars.

One of the key parameters in unraveling the orbital and physical properties of a binary system is that of the system inclination, *i*, in our line of sight. The inclination, of course, can be at any angle on the sky as we see it, but there is the special case when the inclination is nearly 90° and the component stars within the system will begin to undergo mutual eclipses in our line of sight. The eclipse condition, when one star literally blocks the light from the other star in our line of sight will depend upon the sizes of the stars in the system and their orbital separation. The geometry for the onset of eclipses to occur is illustrated in Figure 1.14, and it transpires that if the two stars in the system have radii R_1 and R_2, then eclipses will occur if the inclination is greater than the value defined by $\cos(i) = (R_1 + R_2) / a$. In deriving

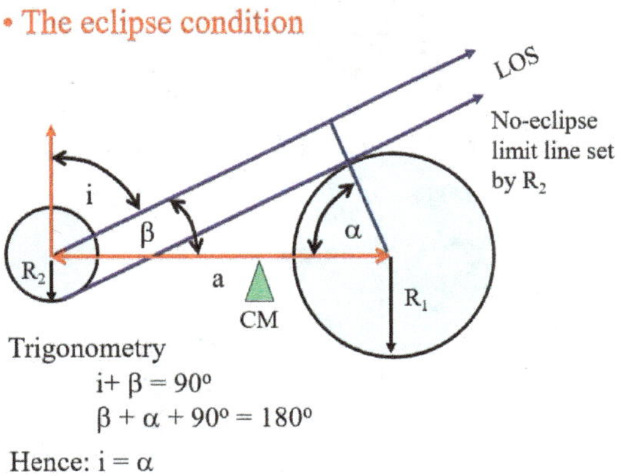

Figure 1.14. The geometry required for the two stars in a binary system to undergo eclipses in our line of sight (LOS). The plane of the orbit is indicated by the horizontal red line connecting the star centers. The vertical red line is at ninety degrees to the plane of the orbit. Here the two stars have radii R_1 and R_2 and the orbital semi-major axis is *a*. The key piece of geometry is to show that angle $\alpha = i$, where *i* is the system inclination in our line of sight.

the critical inclination condition, we have made use of the trigonometric identities: $\sin(\beta) = \sin(90 - \alpha) = \cos(\alpha) = \cos(i)$.

An important subgroup of binary systems are the so-called eclipsing binaries in which our line of sight is directed along the system's orbital plane (i.e., when $i = 90^\circ$). In this case, the two stars will periodically pass one in front of the other as we see them on the sky. At each half orbital period, one star will pass in front of the other as we see the system, and this will cause a periodic variation in the system brightness. This variation can be monitored and displayed according to a light curve, where brightness is plotted against time (Figure 1.15). The eclipse / non-eclipse brightness pattern will repeat according to the system's periodicity, P.

To unravel the observations in terms of the radii of the two stars within an eclipsing system (i.e., as shown in Figure 1.15), we can use the light curve to determine the times T_1 through to T_4 and then consider the physical distance moved by star 2 relative to star 1. In this manner, we see that in the time interval T_1 to T_4, star 2 has moved a distance (in our line of sight) of $2(R_1 + R_2)$. Likewise, in the

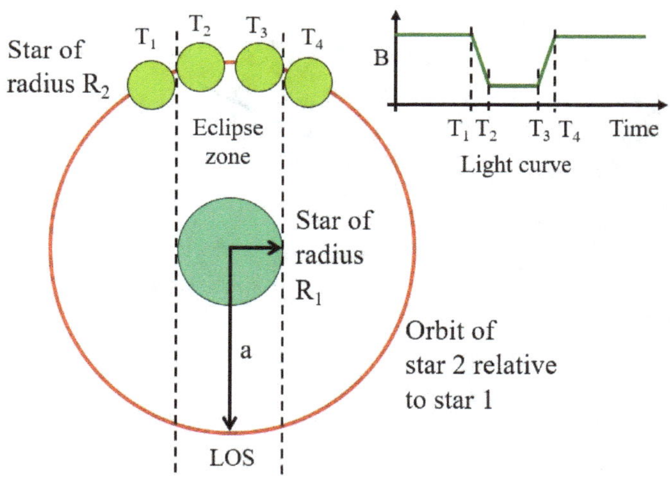

Figure 1.15. A plan view of an eclipsing binary system. The observer is imagined to see the system along its orbital plane in the direction of their line of sight (LOS). The inset at the upper right shows the light curve that will be observed as star 2 passes behind star 1 in the LOS. The eclipse time interval T_1 to T_4 can be measured by monitoring the system's brightness.

time interval T_2 to T_3, star 2 has moved in our line of sight through a distance $2(R_1 - R_2)$. Assuming for the sake of convenience that the binary system has a circular orbit, we can further link the distances moved by star 2 to the system parameters. In our line of sight the constant velocity with which star 2 moves with respect to star 1 will be $V = 2\pi\, a / P$, where a is the orbital separation and P is the orbital period. Since the velocity of star 2 about star 1 is constant, we can construct values for distance traveled over time taken to obtain

$$\frac{2\pi a}{P} = \frac{2(R_1 + R_2)}{T_4 - T_1} = \frac{2(R_1 - R_2)}{T_3 - T_2} \tag{1.16}$$

By measuring the eclipse time intervals (T_1 through T_4), deducing the orbital period, P, and determining the orbital separation, a, with a parallax measurement, there is enough information in Equation (1.16) to deduce the individual values for R_1 and R_2. Since eclipsing binaries are relatively rare, there is not an extensive data set on directly determined stellar radii to work with, but Figure 1.15 shows a sample of those values that have been published for main sequence stars. There is a much greater spread in the mass-radius diagram, at a given mass, than seen in the mass-luminosity diagram (Figure 1.13). This is a result of the relatively large increase in the radius of a star as it ages during the main sequence phase (see Chapter 2). Nonetheless, the lower boundary of data points plotted in Figure 1.16 suggest an approximate mass-radius relationship of the form $R \sim M^{0.8}$ for stars during their early main sequence phase.

By looking at stars located within eclipsing binary systems, measuring the system parallax, and recording the spectra and magnitude for each star, astronomers are able to deduce values for the physical mass, size, luminosity and temperature of the stars. These physical values for main sequence stars turn out to be closely correlated, and in a 3-D diagram of mass versus luminosity and radius, the data points do not fall at random in the diagram, but along a well-defined diagonal line (Figure 1.17). This is a remarkable result, and in the early 20th century, explaining why it comes about was a major problem for astrophysics to solve. In Chapter 2 and 3, we shall look more closely at how the luminosity, temperature and radius change as a star, with a given mass, ages.

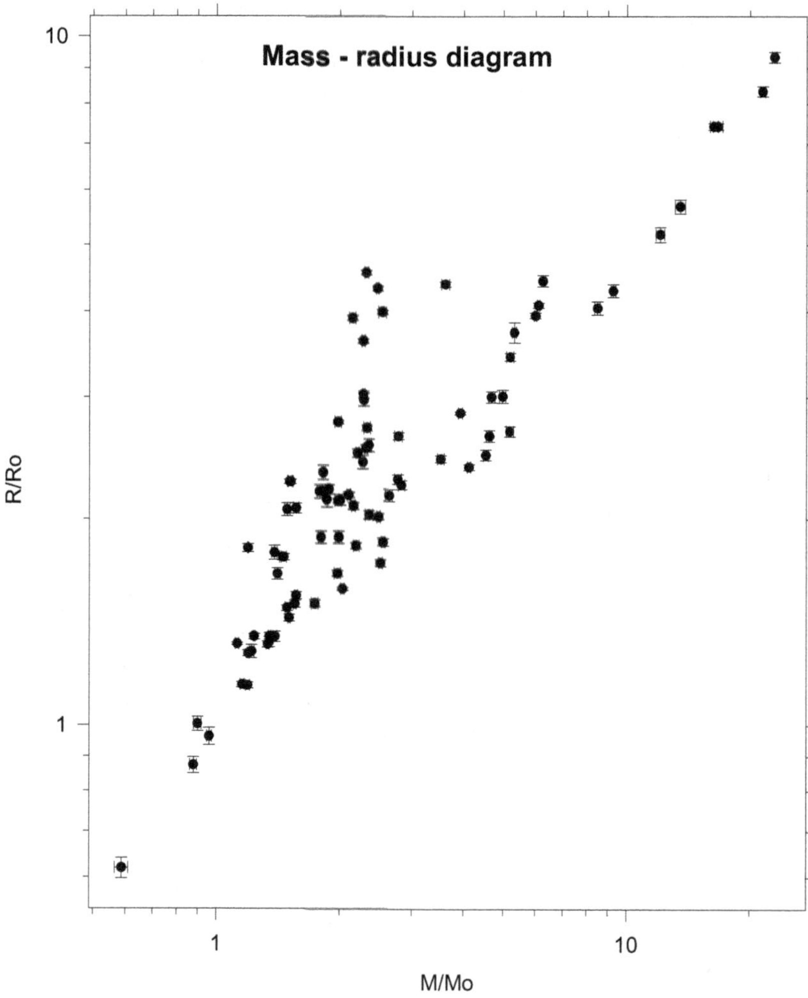

Figure 1.16. The mass-radius diagram for main sequence stars. The spread in the radius at a given mass is due to age variations — older main sequence stars are larger than younger ones. Data from [2].

Determining the formation age of a star is a non-trivial task. In the case of the Sun, its age is estimated by making laboratory-based measurements on meteorite fragments. Such radionuclide ages make the Sun (just like Earth) some 4.56 billion years old. There are means of estimating the age of stars within star clusters (see Chapter 2), but in general, the only way to determine the approximate age of a single

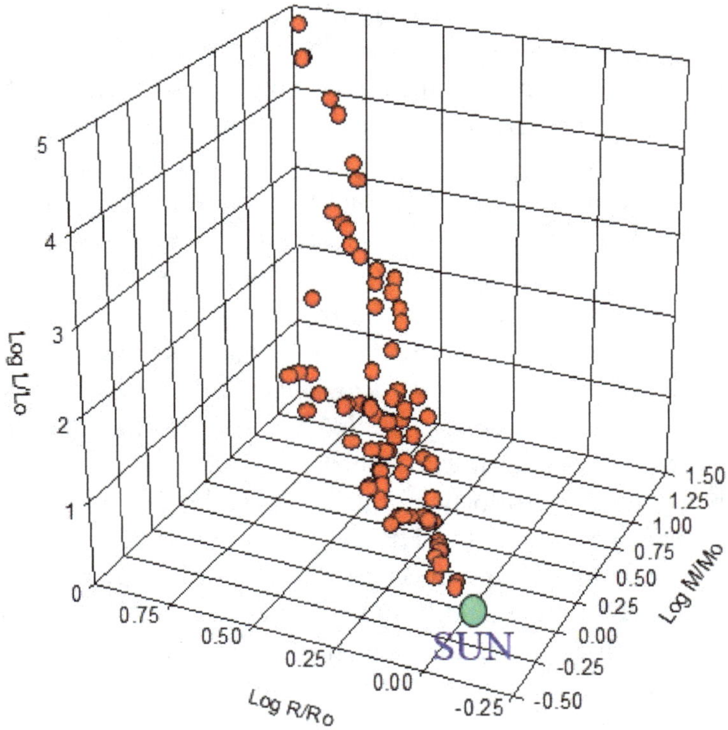

Figure 1.17. The mass-radius-luminosity relationship for main sequence stars. The data points do not plot at random in the diagram but follow a reasonably well-defined diagonal. Data from [2].

star is to estimate its mass, luminosity, and temperature and compare such values to those provided by detailed grids of model calculations derived from stellar evolution computer codes.

1.8. The HR diagram

To Sherlock Holmes she is always the woman
(*A Scandal in Bohemia*)

The Irene Adler-equivalent of astronomy diagrams is the Hertzsprung-Russell (HR) diagram. It is *The Diagram*, the veritable battle ground, no less, in which theory confronts observation, and it is where the primacy of astrophysics is brought to the fore. In essential detail, the

HR diagram is a plot of luminosity against temperature — although it has many proxy forms (specifically, for us, absolute visual magnitude versus color index). Figure 1.18 shows the HR diagram of stars observed with the GAIA spacecraft[30], where the luminosity and temperature (along with spectral type) are shown on the righthand and top axes respectively. On the lefthand and bottom axes, the scales are expressed in terms of absolute magnitude[31] (recall Equation 1.9) and color index[32]. The oddity of the HR diagram is that contrary to the usual numbering convention, the x-axis temperature values decrease as one moves away from the origin tracking from left to right. This is due to the historical legacy of plotting the x-axis according to spectral type before it was realized that the spectral sequence was a temperature sequence.

When Ejnar Hertzsprung, and independently Henry Norris Russell, first constructed the HR diagram circa 1911, they plotted absolute magnitude against spectral type, and found that the majority of stars fell on a diagonal, running from very hot and very luminous stars (with large negative absolute magnitudes and O spectral type) to cool and low luminosity stars (with large positive absolute magnitudes and an M spectral type). This diagonal became known as the main sequence, and indeed, it corresponds to those stars in their main sequence evolutionary phase, The main sequence also delineates the luminosity and temperature variation of luminosity class V stars. For masses larger than about 1.5 $M_\odot$ on the main sequence, the luminosity and temperature vary as a power law of the form $L \sim T^{8.5}$. At lower masses, the slope changes to $L \sim T^{5.5}$. This change in dependency, it turns out, relates to a changeover in the

[30]Launched in 2013, the GAIA spacecraft has been measuring the parallax and proper motions to nearly a billion stars. The GAIA data catalogs provide, accordingly, the detailed positions and orbits to about 1% of the stars within our Milky Way galaxy.

[31]When looking at a cluster of stars, the apparent magnitude can be used on the assumption that the stars are all the same distance. This effectively assumes that the spread, or size of the cluster is small compared to its distance away from us.

[32]The color index is expressed in GAIA specific filters, rather than the B and V filters of the Johnson system, and the calibration with temperature as given in equation 1.12.

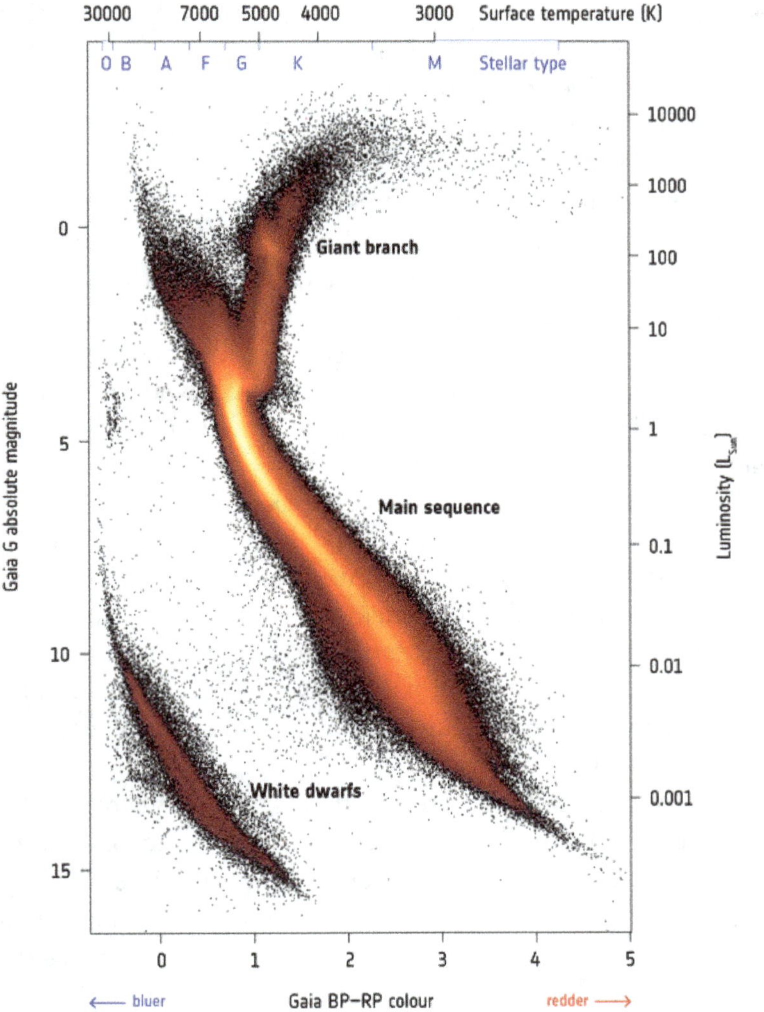

Figure 1.18. The HR diagram derived from GAIA spacecraft data. This image is based upon the observation of 4 million stars located within 5000 light years of the Sun. Image courtesy of ESA and the Gaia Data Processing and Analysis Consortium (DPAC); Carine Babusiaux, IPAG — Université Grenoble Alpes, GEPI — Observatoire de Paris, France.

central fusion reaction sequence responsible for generating internal energy (as will be described in the next chapter).

Given that the HR diagram is effectively a log-log plot of luminosity versus temperature, we can deduce from Equation 1.16 that lines indicating a constant radius R must plot as parallel diagonal lines of slope = 4 running across the diagram. Furthermore, we also see that in the upper righthand corner of the HR diagram, stars will be highly luminous, cool and very large. Whereas in the lower lefthand corner of the HR diagram, stars will have very low luminosities, high surface temperatures and small radii. Accordingly, stars plotted in the upper righthand corner of the HR diagram are called giants, and those in the lower lefthand corner of the diagram are called dwarfs. The dwarf or giant nature of a star being set relative to the size of a star on the main sequence with the same temperature. Figure 1.18 shows a giant branch peeling away from the main sequence, and a white dwarf sequence delineated in its lower lefthand corner. As we shall see in Chapter 2, the stars evolving towards the giant branch, crossing the so-called Hertzsprung Gap, are those stars that have left the main sequence and have yet to initiate helium fusion reactions in their core. White dwarfs, in contrast, are on a cooling sequence and do not generate energy within their interiors. In the solar neighborhood, careful survey analysis indicates that approximately 92% of stars are main sequence stars (luminosity class V), some 7% are white dwarfs, and about 1% are giants (luminosity class I to IV).

In terms of spatial distribution, that is, the number of stars of a given spectral type and luminosity class per cubic parsec, n^*, of space, various surveys provide the data shown in Table 1.3. In terms of main sequence (luminosity class V) stars, the O and B spectral type are relatively rare, with the most common star being that of spectral type M. The volume of space required to contain at least 1 OV star is

$$\frac{4}{3}\pi r^3 n^* = 1 \tag{1.17}$$

From Table 1.3 we have that for OV stars $n^* = 2 \times 10^{-8}$ stars/pc^3, and accordingly the nearest such star to the Sun will be located at a distance in the order of $r = 228$ pc[33]. In this same volume of space,

[33] This number is a typical distance between any given star (not just the Sun) and the next nearest OV star. Accordingly, it is an average separation. It transpires that the nearest O spectral type star to the Sun is Zeta Ophiuchi at $r = 112$ pc.

Table 1.3. Spatial density n^* of stars in the solar neighborhood in terms of spectral type and luminosity class. The units are in number of specific spectral type stars per cubic parsec. The white dwarfs (WD) are placed in the table according to the spectral type equivalent to their surface temperature.

	O	B	A	F	G	K	M
I ⇒ IV				5×10^{-5}	2×10^{-4}	4×10^{-4}	3×10^{-5}
V	2×10^{-8}	1×10^{-4}	5×10^{-4}	3×10^{-3}	6×10^{-3}	1×10^{-2}	5×10^{-2}
WD		1×10^{-3}	2×10^{-3}	1×10^{-3}	6×10^{-4}	1×10^{-3}	

there will be some 298,000 Sun-like stars (spectral type G and luminosity class V), 2.5 million MV stars, some 280,000 white dwarfs and 34,000 giant stars (across all spectral types with luminosity class I to IV). In terms of its numbers, our Milky Way galaxy consists of some 300 billion stars and is predominantly composed of K and M spectral type (red dwarf) stars. It has been estimated that there are in the order of 20,000 O spectral type stars in our entire galaxy, and while the red dwarfs account for the greater number of objects, a glance at the fourth column of Table 1.2 indicates that it is the relatively few O spectral type stars that dominate with respect to the energy being radiated by main sequence stars into space per unit time.

It is a notable quirk of our present-day existence that not one of the most common kind of stars within the Milky Way galaxy (MV spectral type) is actually visible to the naked-eye. Is this, one can question, a reasonable situation given what we know about stars? The data in Table 1.3 indicates that the typical volume of space required to find at least one MV spectral type star is about 20 pc^3, given a typical distance of $r = 1.68$ pc. Table 1.2 further informs us that the absolute magnitude of the brightest MV spectral type star is that corresponding to an M0V star with $M_V = 8.9$. Using this magnitude and a characteristic separation of 1.68 pc, Equation (1.17) indicates an apparent magnitude of $m = V = 5.0$. Technically, such an apparent magnitude is within the realm of naked eye detection. The closest M0V star to the Sun, however, is Lacaille 8760 (AX Microscopii) with a distance of 3.87 pc, and an apparent magnitude of 6.67 that is below the detection limit for the human eye. In fact, the closest star to the Sun is an MV star — this being Proxima

Centauri. Designated an M5 V star, Proxima has, according to Table 1.2, an absolute magnitude of 14.5. Combining this with its known parallax distance of 1.3 pc, Equation (1.19) indicates an apparent magnitude of 10 — well below the threshold for naked eye detection. Since stars are in relative motion, however, sooner or later a close encounter with the Sun can will occur. For an M0 V star to become visible to the naked eye, with a limiting apparent magnitude threshold of $m = 6$, it will have to pass within 2.63 pc of the Sun, and this is something that is likely to happen at intervals in the order of 22,500 years (see Chapter 3 for further discussions).

1.9. Star clusters

To the naked eye, it is not immediately obvious that stars can cluster together. Indeed, to the eye they appear relatively random in their distribution across the heavenly vault. For sure, structure can be seen through the smallest of telescopes; the fuzzy luminosity of the Milky Way, the faint glow in the constellation of Andromeda[34], and the luminous haze in the Pleiades cluster make them stand out from the background. Likewise dark clouds, void regions apparently empty of stars, can be seen. But these regions, while intriguing, are, to the eye's ability to resolve, distinct from the pointillist scattering of the stars on the sky.

> *Queer certainly! But what do you gather from it?*
>
> (*The Valley of Fear*)

Of the non-star-like features visible in the heavens, the diaphanous arch of the Milky Way is the most distinctive and diagnostic. Indeed, the Milky Way, as revealed by telescopic observations, is composed of stars, myriads of faint and distant stars, and its shape informs us that they have a flattened, disc-like, spatial distribution. The Sun and solar system are embedded within the Milky Way, and

[34]This, in fact, is the most distant object that the unaided human eye can discern and it corresponds to the Andromeda galaxy, which is located some 152,000 light years away.

when we look in the direction of Sagittarius, we are gazing directly towards its central core — what we call the galactic center. The idea that the Sun is situated within the disc of the galaxy, and that the universe is sprinkled with distinct clusters of galaxies, is a result of recent genesis in the long history of astronomy. That galaxies are both large and remote from each other only became clear in the 1920s and 30s, and that galaxies form clusters (and super clusters) is a result of the 1970s and 80s.

The basic halo, central bulge, and disc-like structure of our Milky Way galaxy was illustrated earlier in Figure 1.3, and while the galactic halo contains globular clusters, the galactic disc contains open clusters. Of the thousand or so open clusters known in the Milky Way galaxy, the Pleiades (Figure 1.19) represents one of the closest such structures to the Sun, being located some 136 pc away.

Figure 1.19. Hubble Space Telescope image of the Pleiades Cluster. The blue coloration of the image is related to the preferential wavelength scattering of blue light by interstellar dust. This galactic cluster is visible to the naked eye. There are no known (or, at least recognized so far) blue stragglers within the Pleiades. Image courtesy of ESA.

This cluster is relatively young, being an estimated 110 million years old[35], and it contains at least one thousand stars, although only the brighter ones are obvious to the human eye. In general, open clusters live up to their name; they have a loose, open structure and are found exclusively in the discs and arms of spiral galaxies. While having more stars per unit volume of space than the solar neighborhood, with n^*(cluster) ~ $10n^*$(solar neighborhood), open clusters are not strongly gravitationally bound structures, and they gradually lose stars to the surrounding disc of the galaxy. It is estimated, for example, that the stars within the Pleiades will have all dispersed (that is, 'evaporated' into the galactic disc) within some 250 million years from now.

While 'Shine bright, live fast, and travel far' is a motto that might well be applied to the stars in an open cluster, that for stars within a globular cluster might be 'live long and keep your companions close'. The globular clusters associated with our Milky Way galaxy are characteristically found above and below the galactic disc, distributed like a swarm of bees about a central galactic core. About 150 globular clusters are presently known, and their name describes their appearance. They are generally spherical in appearance and may contain many hundreds of thousands to many millions of stars (Figure 1.20). A few globular clusters are visible to the naked eye, although they have a star-like appearance — perhaps the best such sightings are those of Messier 13 in the constellation of Hercules, and Omega Centauri in the constellation of Centaurus. The spatial distribution of stars in a globular cluster increases inwards, being characteristically about 0.5 stars per cubic parsec in the outermost regions and some 100 to 1,000 stars per cubic parsec in their inner cores. As will be seen, this dramatic increase in the number of stars per unit volume of space has important consequences for the emergence of blue stragglers (as well as other exotic star systems[36]).

---

[35] This age is based upon matching measured star characteristics to evolutionary computer models of stellar structure and ageing (see Chapter 2).

[36] E.g., millisecond pulsars, and low-mass X-ray binaries, and even, as in the cases of at least Omega Centauri, and M 15, intermediate mass black holes (see Chapter 2).

Figure 1.20. The globular cluster Omega Centauri. Located some 5,240 parsecs from the Sun, the cluster is some 11.5 billion years old, and contains in the order of 10 million stars, of which some 50 are presently recognized as being blue stragglers. The cluster has a total mass of about 4 million Suns, and is some 570 parsecs across. Image captured with the WFI camera from ESO's La Silla Observatory.

The age of a star, as indicated before, is a very difficult quantity to gauge, but in the case of stars within a cluster, one can determine the age of the cluster itself through astrophysical model comparisons, and the assumption (a good one as it turns out) that all the stars in the cluster formed at essentially the same time. It will be seen in Chapter 2 that the manner in which a star ages is critically dependent upon its initial mass, and studies of star formation regions indicate that stars within a cluster generally form with a whole range of masses — typically between about 0.1 to 100 times the mass of the Sun, with the most common mass being about ½ that of the Sun. Furthermore, in Chapter 2 we shall also see that the more massive a star is, the more rapidly it evolves through its main sequence phase to become a red giant. With these caveats in place, the age of a cluster can be estimated according to the precept that all the stars within it formed at the same time, have the same composition, and formed with a wide range of stellar masses. From this, stellar models, over a range of different stellar masses, can be constructed and the luminosity, temperature and radius of the different mass stars can be compared at a specified age — producing what are called stellar isochrones. The cluster age is determined by fitting isochrones to the main sequence stars delineated in its specific HR (color-magnitude) diagram.

In terms of their dynamics, the chief difference between open clusters and globular clusters is that the former orbit the galactic center along circular orbits coincident with the galactic disc, while the latter, in contrast, orbit the galactic center along highly elliptical orbits that are inclined at random angles to the galactic disc. In terms of their stellar make-up, the open clusters tend to contain relatively young, massive stars with metallicities similar to, or even higher than, that of the Sun; globular clusters, in stark contrast, contain, old, low mass stars with metallicities many times lower than that deduced for the Sun. Indeed, globular clusters contain near primordial stars, placing them among the first star systems to have formed some 12.5 billion years ago, when the universe itself was less than a billion years old.

1.10. Placing the Sun

What the eye rests upon, the mind dwells upon
(Arthur Conan Doyle — *The Narrative of John Smith*)

When Copernicus, in 1543, published his *De Revolutionibus*, he controversially placed the Sun at the center of the universe and made the Earth move along the third orbit outwards. This was a revolutionary (although not totally new) idea, and while it eventually changed humanity's outlook concerning the structure of the heavens, it rankled some observers in that it removed humankind from the center of importance. This idea of removal was latter expressed in terms of the Copernican Principle, in which it is argued that observations made from the Earth will be the same as those that might be made from any other randomly selected position in the Universe. The position of the Sun, solar system and us (as living beings on the Earth) are not, however, random and/or even average occurrences. Indeed, the space and time location of the Sun within the galaxy are decidedly special, and we are a direct result of the special conditions needed to support life. For all this, while Copernicus was incorrect to place the Sun at the center of the Universe (indeed, the Universe has no specific center), it was only in the second decade of the 20th century that astronomers realized (or rather, found through observations) that the Sun was not at the center of the Milky Way galaxy.

Key to placing the Sun's location within the galactic disc and determining its distance from the galactic center was the unraveling of the observed distribution of globular clusters. What is needed, it turns out, to understand the distribution of globular clusters is a catalog of observations providing their sky locations and distances. This was a project undertaken by astronomer Harlow Shapley, and he specifically used the detection of RR Lyrae stars located within globular clusters to determine their distances. Recall, RR Lyrae stars are pulsating stars with a distinctive variation in brightness (Figure 1.12), and these stars have a known (that is, calibrated) absolute magnitude (or luminosity — see Figure 1.11). Accordingly, a measurement of RR Lyrae star with apparent V magnitudes will enable a distance estimate to the host globular cluster to be made (as per Equation 1.10). Combining the distance and spatial direction observations, under the assumption that globular clusters were distributed in a spherical manner about the galactic center, Shapley was able to show that the center of the galaxy was located in the direction of the constellation of Sagittarius (as seen from Earth — see Figure 1.21), and that the Sun (using modern determinations) is located some 8,000 pc from the galactic center. This relative remoteness of the Sun, situated about halfway out in the galactic disc, is important for our existence. If the Sun was much closer to the galactic center, then, at earlier times, the higher supernova and gamma-ray burster rates experienced at this closer position would have

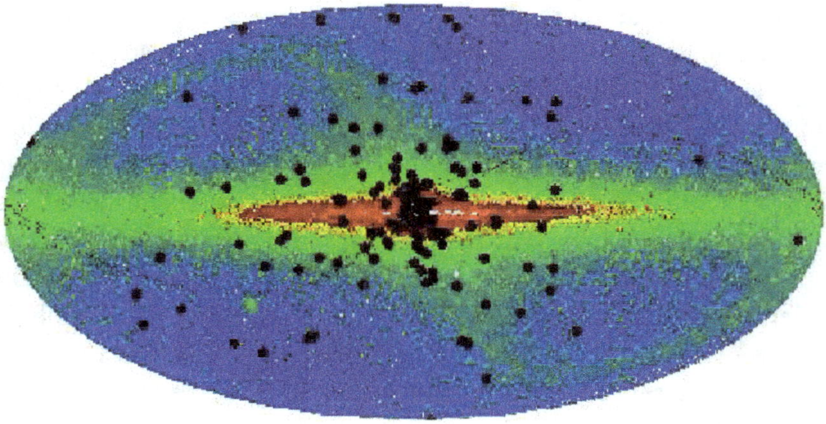

Figure 1.21. The spatial distribution of globular cluster about the center of our Milky Way galaxy. The clusters are plotted as black dots set atop an all-sky map of the 2.2-micron, thermal emission from galactic dust. The background image was obtained by instruments carried aboard the COBE spacecraft. The central, horizontal, green band corresponds to dust emission in the galactic disk, while the yellow, red and white regions indicate intense dust emission from the direction of the galactic center. The faint, reverse-S-shaped emission (seen in green) corresponds to thermal emission from dust within the ecliptic plane of the solar system. In this image, it is clear that most of the globular clusters are found in the direction of the galactic center, indicating that we must be displaced away from the core of the galaxy [3].

potentially sterilized any newly formed life on a young Earth. Being further out from the galactic center, however, allowed for a less disruptive environment in which life could evolve. Furthermore, since the solar system formed many billions of years after the formation of the galaxy[37], this allowed for a metal-rich interstellar medium to exist. This latter point is important since it is from the 'metals', as such, that planets and life (including human beings) are composed. Our galactic position is not just due to random circumstance, but is specific and privileged with respect to our existence.

[37]The solar system is some 4.56 billion years old, while the galaxy is about 13.6 billion years old.

1.11. Climbing the cosmic ladder

The marks of the ladder attract[ed] Holmes's special attention
(Arthur Conan Doyle — *The Adventure of the Tall Man*)

While the distinct properties of RR Lyrae stars enable a distance measure of globular clusters to be made, the much more luminous Cepheid variables (recall Figure 1.11) can be used to find the distances to other galaxies. Indeed, these variable stars, and their calibrated properties, are taking us several rungs further up the cosmic distance ladder (recall Figure 1.6). Here is the power of local calibration. If we know what stars are doing in our cosmic backyard, so to speak, then by assuming that they will do the same things in any other backyard, in any other galaxy within the Universe, then we can literally measure the large-scale distribution of matter. This is a powerful and incredible result, and provided our standard candles remain true, over both cosmic time and space, then we do have the power to know and measure the Universe (and all this without leaving the comfort of planet Earth).

In the timespan of just a few hundred years, astronomers have moved from accurately measuring the size of the Earth's orbit (the astronomical unit) for the first time, to measuring the distances to individual stars (parsecs away), to probing the scale of star clusters and our Milky Way galaxy, to determining the large-scale structure of the Universe with the detection of voids, clusters, and superclusters of galaxies. Indeed, while astronomers and philosophers have been trying to understand the Universe since time immortal, we are only just one century into the era of modern cosmology, where actual measurement data exists. There is no precise beginning, or single pioneer who set the ball of modern cosmology rolling, but certainly the introduction of general relativity by Albert Einstein in 1915 marks a reasonable starting epoch. Prior to this, however, Vesto Slipher, in 1912, had, for the first time, measured Doppler shifts (recall Equation 1.2) in the absorption lines of several galaxies, and discovered that they were travelling at vast speeds through space — some towards us and some away. The absorption lines detected in the spectrum of the Andromeda galaxy, for example, were blue shifted and indicated that it was moving towards us at an apparent speed of about 300 km/s. By 1915, Slipher had measured more galaxy speeds and found that the majority were actually moving away from us — the absorption lines were red shifted, indicating a positive velocity of motion away from us. While Slipher continued to measure the speeds with which galaxies were moving, Edwin Hubble (along with many other astronomers in the mid-1920s) began determining the distances to galaxies. Hubble used observations of Cepheid variables, and standard ruler techniques (recall Figure 1.6) to determine distance. Then, in 1929, Hubble published his famous result that indicates the existence of a velocity–distance law for galaxies. Known as Hubble's Law, this result states that if the recession velocity of a galaxy is V (km/s) and the distance is D (Mpc), then

$$V(\text{km/s}) = \mathrm{H}_0\, D(\text{Mpc}) \tag{1.18}$$

where H_0 is Hubble's constant. This remarkably simple relationship tells us something distinctly profound: that the Universe is expanding uniformly. That is, as time goes by, the Universe grows steadily in size, and equally in all directions. This observational result was soon connected with the earlier theoretical ideas (based upon the general theory of relativity) published by Alexander Friedmann in 1922. The theories and the observations were clear, the Universe is expanding, and this implied the incredible result that if one traced the motion of the galaxies backward in time then the universe must have had a moment of creation — this still-mysterious, coming-into-being moment, is now called the eponymous "Big Bang". This is again an incredible result and tells us that the observable Universe is not infinitely old, as many had previously thought (or, rather assumed), but it burst into existence at a definite time set in the finite past. Astronomers and theoretical physicists have been trying to find a physical understanding and description of this moment of origin ever since. They are still trying.

Not only does the reach of astronomy now extend over many gigaparsecs in distance, it also extends over many billions of years in time. Indeed, the Universe is estimated to be some 13.8 billion years old, and remarkably, we can readily see far back in time to when the Universe was just 378,000 years old. This latter, remarkably precise number, takes us to the time when the cosmic microwave background (CMB) radiation was imprinted upon our view of the heavens. This moment corresponds to the epoch of recombination, a misnomer applied to the time when the Universe was cool enough for free electrons to be captured by free protons to form neutral hydrogen atoms. Prior to this time, the Universe was too hot for neutral hydrogen to exist, and space was filled with a fog of photons interacting with and being scattered by the abundant free electrons. The CMB forms the most perfect blackbody radiator that we know of in nature, with a temperature of $T = 2.72548 \pm 0.00057$ K. The spectrum of the CMB is a near, but not quite, perfect Planck curve, and this is important to our very existence. Indeed, the fluctuations seen in the observed CMB flux compared with expectations from Planck's formula — Equation (1.5) — are equivalent to temperature variations of just a few thousands of a degree Kelvin (Figure 1.22).

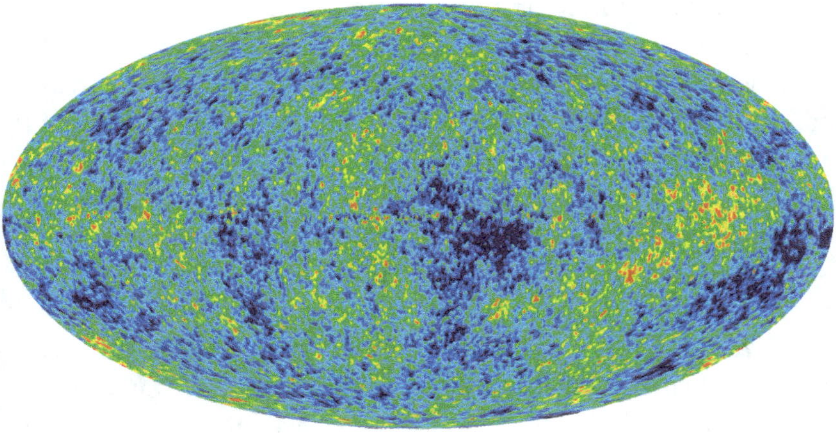

Figure 1.22. All-sky map of the cosmic microwave background radiation. What we are seeing are minuscule temperature variations (shown as color differences) in the early Universe, at a time when it was just 378,000 years old. The temperature range displayed (from red to dark blue) corresponds to just ± 200 microKelvin. Image courtesy of NASA and the WMAP Science Team.

These miniscule temperature variations are important, however, since they indicate regions of slightly higher and/or slightly lower density, and it is from these regions that all of the large-scale structure in the Universe eventually evolved.

Theoretical models take us much further back in time than the epoch of recombination, but currently, there are no instruments capable of measuring what might have happened. In principle, for example, there is an earlier all sky background epoch (when the Universe was just a second or so old) when neutrinos could first fly freely through space, but we do not currently have a telescope that can map this structure out. Even this epoch is ancient with respect to theoretical models of the Universe and its formative moments — these models, some more robust than others, take us back to a time when the universe was a mere 10^{-34} seconds old. But we are now moving too far afield from our main topic of interest. The question for us is when did objects we recognize as stars first begin to form?

The very first stars (Population III stars — recall Section 1.2) appeared when the Universe was about 100 million years old. These stars, however, were unlike any star that can exist in our Milky Way galaxy today. First of all, Population III stars had no metals — they were composed entirely of hydrogen and helium (with a smattering of boron and beryllium) — and they were very massive. Without the aid of metals to cool the first cosmic clouds of hydrogen and helium, star formation resulted in the production of stars with masses in the range of several hundreds to several thousands of solar masses[38]. These stars were highly luminous and they lived for perhaps a few hundred thousand years. Their end phases were either a supernova disruption or a direct collapse into a black hole. Through supernova disruptions, the first elements beyond hydrogen and helium were synthesized and injected into the proto-galactic clouds of gas, adding the small metal abundances seen in old Population II stars. The collapse into black holes fed into the intermediate mass black hole

[38] Today, the reverse is true, and the conditions in star forming regions now favor the formation of low mass stars.

population, and these objects potentially acted as seed centers for the first-formed globular clusters. Galaxies, as vast assemblages of gas, dust, and stars, began to proliferate from about 1 billion years after the Big Bang.

1.12. The oddballs

Behold the fruit of pensive nights and laborious day
(*His Last Bow*)

This chapter has covered an immense amount of cosmic time and space. We have, in the process, attempted to unravel the history of stars, from the first Population III stars, formed some 13.5 billion years ago, to the Sun, formed some 4.56 billion years ago. Indeed, over the past several centuries, astronomers have learned how to apply basic geometrical ideas to determine the distances to the stars (and the greater Universe beyond), and have applied fundamental physical theory to determine what stars are and how they function. Indeed, what this first chapter has been all about is the discovery of stars: the very who, what, where, when, and how of their existence — there is no specific why. Remarkably, for it is remarkable, by measuring periodic shifts in their positions, deducing velocities, recoding energy fluxes, and applying a smattering of physics and chemistry, the stars can be understood. And, remarkably again, it is fair to say that we now know an awful lot about stellar constitution and evolution. They are physical objects with a mass, radius, surface temperature, and luminosity. All these physical quantities can be measured in one way or another, and during the past century and a half, astronomers, through long and hard graft, have enumerated the properties and distribution of the stars in ever increasing detail. We know the stars, and from this knowledge we can construct a typical picture — a normalcy — of what stars are. There are, of course, many details that have yet to be decided upon, and there are many parameters that require additional refinement, but the basics so far deduced will not change in any substantial manner as we move into the future. The same cannot be said of our understanding of large-scale cosmic structure, and of how the Universe itself came into existence.

While astronomers, in general, will continue to pursue the fundamental work needed to finesse our observational understanding of star basics, what often catches the eye and interests of researchers are the oddities — literally the stars that stand out from the crowd. These oddities are the stars that defy placement in the standard classification schemes, or have characteristics suggestive of some atypical set of formation circumstances; these are the stars that catch hold of the imagination. Indeed, this is the case with the blue stragglers. We can only deduce their existence as special objects, however, by knowing the behavior and characteristics of 'ordinary' stars, and 'ordinary' star systems; blue stragglers, it turns out, are equivalent to the individual lost within a crowd.

> *Education never ends, Watson. It is a series of lessons with the greatest for the last*
>
> (*The Red Circle*)

I think Holmes[39], in spite of being a hard critic and exacting logician, would be genuinely impressed with what modern astronomers have been able to achieve with respect to the study of stars, and the annotation of their physical properties. The practitioners of astronomy have built a great chain of reasoning, interspersed with leaps of human genius and careful observation. Indeed, although no astronomer himself (at least, according to Watson), Holmes would have thoroughly approved of the methods deployed by the astrophysicists. They took the available data, they refined and defined it, they mulled over its meaning, and following sound theoretical and philosophical reasoning, they obtained a measure of the stars. It is no small thing that this information is available to us in such intricate and intimate detail, and we should reflect and marvel that we know so much about such distant and remote objects as the stars. For all this, how-

[39] I am here buying into the notion that he was a real person.

ever, our knowledge is incomplete, and from the great depths of our accumulated understandings new questions arise, and these questions require new and informed answers. Indeed, to have a mystery, as in the case of the blue stragglers, we need a normalcy against which to see them contrasted, and by understanding the blue stragglers we have the promise of obtaining a new and more refined normalcy.

> *There, Watson! What do you think of pure reason and its fruits?*
> (*The Valley of Fear*)

References

[1] M. Beech. *Alpha Centauri: unveiling the secrets of our nearest stellar neighbor*. Springer, New York, 2015.

[2] Torres, G., Andersen, J., and Giménez, A. Accurate masses and radii of normal stars: modern results and applications. *Astronomy and Astrophysics Review* 18, 67–126 (2010). https://doi.org/10.1007/s00159-009-0025-1.

[3] Living reviews: relativistic binaries in globular clusters — by Matthew J. Benacquista and Jonathan M. B. Downing. https://www.emis.de/journals/LRG/Articles/lrr-2013-4/articlese2.html.

Chapter 2

How the stars work

In *The Illustrious Client*, Sherlock Holmes explains to Watson that, "I am accustomed to have a mystery at one end of my cases, but to have it at both ends is too confusing." In the same manner, and in order to gain some knowledge of what it is that might turn a star into a blue straggler, we must first know what it is that makes an ordinary star work and function. In reality, of course, there is no such thing as an ordinary star. There are classes of stars, but each star has its own set of peculiarities and properties that make it unique. The Sun, for example, is a unique and entirely uncommon star within the Milky Way galaxy [1]. It is so since we live on a planet that orbits about it. This very fact makes it special not only to us, but to any other observer anywhere else in the Universe. Holmes reminds us in *The Red-Headed League* (with a nod, in the modern era, towards the search for extraterrestrial intelligence — see Chapter 4), however, that, "As a rule, the more bizarre a thing is, the less mysterious it proves to be." For all this, the Sun is a rare type of star; it is not at all a typical star within the Milky Way galaxy in either a group sense (that is, as a member of the G-spectral type class of stars), nor in its planetary-system–supporting sense. Furthermore, in the case of the Sun, we do not fully understand the workings of its outermost layers, nor the internal machinations that control its magnetic and sunspot cycles. For all that, astronomers consider the Sun, and stars in general, as well understood objects. In this sense, they are talking about the gross operating characteristics of stars rather than their individual quirks and variations[1]. In terms of their general astrophysical properties, it is reasonably well known what makes a star tick, although there are still numerous physical details that have, as yet, not been fully explained and/or understood. One can certainly make a detailed computer model of a star; such models will provide a good representation of what stars, in general, are found to be doing, but, for all that, all computer models rely upon the use of simplified physical expressions and special parameterizations. Astrophysics provides us with an essential understanding of the stars, but the Universe has found numerous ways of making the general picture of a star much more object-specific, and much more complicated than astrophysicists typically let on. All this, of course, is what makes the astrophysics of stars so interesting.

2.1. The life of a star

> *Since I first took to milk, I have had it driven into me that the Sun was an enormous mass of burning matter*
>
> (Arthur Colan Doyle — *The Narrative of John Smith*)

Renowned, if not opinionated and controversial, astrophysicists Fred Hoyle once remarked at a conference that stars are basically simple objects. To this point, a member in the audience returned the comment, "You'd look simple from ten parsecs, Fred." In a seemingly contradictory sense, both Hoyle and the audience member are right. Stars are indeed simple objects; they are essentially large, spherical masses of gas slowly collapsing under gravity. They are also immensely complex physical structures powered by the subtle

[1] An example of an apparently unique occurrence (that is, of course, until a second such example is observed) is that associated with Tabby's Star. This system (KIC 8462852), contains an apparently 'normal' F spectral type, main sequence star, but shows dramatic and unaccounted for variations in brightness. Of course, 'quirks and variations' are just terms used until such observations are understood, and one assumes by ordinary, rather than fantastic, operating mechanisms.

machinations of quantum mechanics, and the fusion transmutation of atomic nuclei under blisteringly hot, high-density conditions.

It was recognized well over a century ago that the driving force behind star formation and stellar structure is gravity. Indeed, stellar structure is all about the ability of a star to hold off gravitational collapse. If there were no counter interactions taking place within a star, then the collapse beginning at the moment of initial formation would continue unimpeded and result in the formation of a black hole. Indeed, this is what gravity is continually trying to do to a star: it is trying to drive the collapse to an ultimate end singularity, and thereby produce a final black hole state. Collapse, however, is impeded through the emergence of various internal energy sources, and, up to a point, that is up to some limiting mass M_{max}, the final resting state of a star is governed by the existence of a quantum mechanical exclusion principle acting upon either electrons (in the case of white dwarfs) or neutrons (in the case of neutron stars). The 'up to a point' condition is determined by the star formation process and the chance determination of a star's initial mass. The most massive stars end their days by undergoing catastrophic collapse, forming either a remnant neutron star or, if they are particularly massive, a remnant black hole. Figure 2.1 shows the schematic evolutionary pathways for single stars according to their initial mass $M_{initial}$.

For all of the apparent structure and simplicity of Figure 2.1, there are many subtleties and complex details that require explanation. There are various minimum and maximum masses that need to be explained, along with the physical details pertaining to the energy generation and mass loss phases.

2.2. Making a star

It is ... hoped that the reader will find satisfaction in the solution of the problems [of astrophysics] such as [they] may find in a good detective story in which cluses and arguments are fairly presented so as to lead to convincing conclusions.

(W. H. McCrea. *Physics of the Sun and Stars.* Hutchinson's University Library, London, 1950)

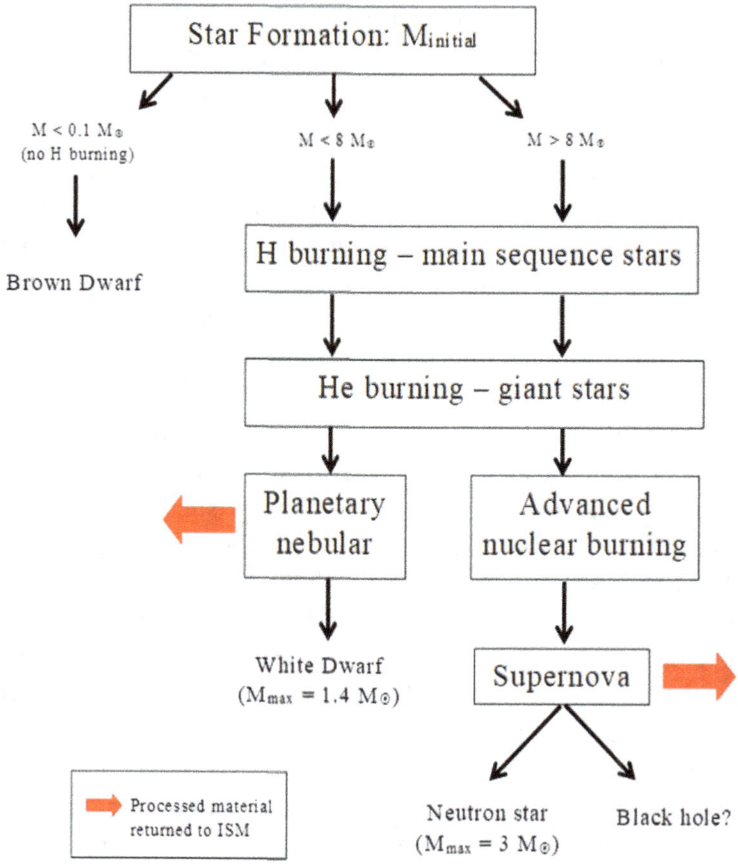

Figure 2.1. Evolutionary channels for single stars of (from left to tight) increasing mass. The solid red arrows indicate phases where significant surface mass loss takes place. See text for details.

When McCrea was writing his book *Physics of the Sun and Stars* in 1950 [2], electronic computers were in their lumbering infancy, and it was still to be several decades before they became a tool readily accessible to the astrophysicist. Classic texts on the structure of stars had, however, been written long before this time, and one can readily single out Arthur Eddington's, *The Internal Constitution of the Stars* (Cambridge University Press, 1926), and S. Chandrasekhar's, *An Introduction to the Study of Stellar Structure* (University of Chicago Press, 1939). These latter texts, however, were largely concerned with identifying the physics appropriate to describe stellar interiors and numerical approximations that enabled solutions of the equations of stellar structure to be found. As Fred Hoyle was to later remark of these times, (the 1950s and 60s), the key to making progress typically related to how much physics to leave out, and/or attempt to approximate, so that the equations of stellar structure could be solved in some manageable fashion. If too much physical detail was retained, being more honest, as it were, to the reality of the problem, then one invariably suffered from equations that were (then) to difficult to solve by hand calculation[2]. By the 1950s, however, astrophysicists were beginning to move beyond the problem of finding static solutions to the equations of stellar structure, to asking questions about the time evolution of stars. How did they form, and what physical processes controlled their subsequent changes. McCrea sums up the situation nicely, "Suppose there is isolated in space a quantity of matter of mass *M*. It will pull itself together by its own gravitation and form something that can be called, ... a 'star'... some of its gravitational energy will be converted into heat energy and some of this will be radiated away... If nothing happens to stop it, this gravitational contraction will go on... sooner or later, sub-atomic reactions must come into operation... There must come a stage when the rate of release [of sub-atomic energy] is exactly equal to the rate of radiation from the surface... When it is reached, the 'star' will steady up into what can be regarded as an equilibrium state [2]."

Let us follow through the same steps and arguments as McCrea, adding in, wherever useful, some additional details[3]. Stars form in those regions of the interstellar medium where gravity wins out over

---

[2] Here the term 'hand calculation' literally means by hand, pen and paper, or by the hand pulling the crank-arm of a mechanical calculator.

[3] McCrea actually reviewed Chandrasekhar's book in the 22 July, 1939 issue of the journal *Nature*. He writes of the book that, "It constitutes in fact a courageous attempt to reduce a branch of astrophysics to analytical or 'rational' form after the manner of classical mathematical physics." While it is now clear that no such 'exact mathematical' reduction is possible, it is certainly the case that considerable progress has been made in the understanding of stellar structure and in the machinations of stellar evolution. Which is not to say that all the details are now known.

all those other forces, turbulence, magnetic interactions, and thermal motion, that would otherwise prevent collapse from taking place. It is a hit and miss process where the conditions of enough mass being brought together (by chance) in the right location at the right time under the right conditions. Making stars is a dynamical process spread over an extended region of interstellar space, perhaps some ten thousand astronomical units across to begin with, but once it starts, once gravity has set its teeth into the natal gas cloud, then the assembly process is rapid and inevitable. At this stage the cloud is undergoing dynamical collapse and the characteristic timescale for this, τ_{dyn}, can be written in terms of the characteristic density ρ of material, with

$$\tau_{dyn} = \frac{1}{\sqrt{G\rho}} \tag{2.1}$$

where G is the gravitational constant. The low density of the interstellar medium ($\rho \sim 10^{-17}$ kg/m^3) dictates that the collapse time to form a protostar is a few thousands of years, and after this time has elapsed the material that is going to find its way into a star is compressed into a region that is just a few hundreds of astronomical units across.

While the collapse of a gas cloud leading to the formation of a star is driven by the pull of gravity, the collapse is modified by rotation. The effect of rotation is to change the collapse from being radial, with all the material moving directly towards the cloud center, thereby building up a more and more massive central concentration of material, into one in which surrounding the central mass is a disc of rotating material. Even with rotation, most of the mass will still end up in the central protostar, but the disk modifies how the accretion proceeds, and it provides a natural location within which planets can form. The rotating disc also allows for the winding-up of magnetic field lines, producing, thereby, a system of polar jets aligned with the spin-axis of the newly forming protostar. Some of the material funneling in through the disc is accordingly ejected back into space along these polar jets. In this manner, the accretion of material by the central protostar is not one-hundred percent efficient, and the star-formation process is itself disruptive of the region out of which it formed. This later effect can either enhance or inhibit star formation in the greater enveloping interstellar cloud, but for all that, the observations do indicate that stars tend to form in clusters rather than as isolated objects.

The range of masses produced in the star formation process varies from about $1/10^{th}$ that of the Sun, up to about 100 times the mass of the Sun. Nature, however, seems to prefer to make lower mass stars, with the most common mass being about ½ that of the Sun. Massive stars are much less commonly formed, and are relatively rare objects within the galaxy (recall Table 1.3). This being said, while most of the mass of our galaxy resides in the myriad number of its low mass stars, it is the more massive stars that dominate with respect to its total luminosity.

2.3. Making energy

Something unknown is doing we don't know what

(A. S. Eddington)

If stars were simply hot solid objects, then the fact that they radiate energy into space from their surfaces (expressed according to their luminosity) would dictate a gradual cooling-off over time and a concomitant drop in luminosity. This cooling-off would be no different in principle to that of a hot coffee cup, placed upon a table in a large room, cooling off. The fact that the room is cooler than the freshly-brewed coffee dictates that heat energy will flow from the surface of the coffee cup into the greater surroundings, thereby slightly warming the air in the room (and the table), with the result that the thermal energy of the hot coffee liquid is reduced. Eventually a temperature equilibrium will come about with the coffee having exactly the same temperature as the air in the room. Accordingly, if stars were solid and simply cooling off, their final temperature (admittedly after a very long time) would be that of the microwave background: about or below 2 Kelvin (–271 °C). Furthermore, as our solid star slowly cools off, its luminosity will decrease — this follows directly from the Stefan-Boltzmann law (recall Equation 1.6). The end fate of our supposed solid star would be a (near) zero temperature, (near) zero luminosity object — a dire star to say the least, although 'star' (at this stage) is no longer a reasonable description of what we are dealing with. Fortunately, however, stars are not solid objects, and accordingly they have options to keep their interiors warm and remain luminous for extremely long periods of time.

In the same way that gravity is the midwife of star birth, transforming a large, low-density, region of an interstellar gas cloud into a small, high-density proto-star, shrinkage is a means by which a star can generate energy. The gravitational binding energy, U, associated with a sphere of mass M, and radius R is given by the quantity $U = -k\,G\,M^2/R$, where G is the universal gravitational constant, and k is a constant set according to the distribution of matter within the sphere[4]. The equation for U indicates that it becomes larger (and more negative) as R becomes smaller, and accordingly, we now have a way in which a star can maintain its luminosity L. Since the luminosity radiated into space per unit time amounts to a loss of energy from the star, this energy loss can be maintained by the star actually shrinking. Assuming (at this stage) that there are no additional internal energy sources that a star can tap into, the luminosity can be accounted for in terms of a change in the gravitational binding energy. More formally, we can write that the luminosity is determined by the rate of change in the gravitational binding energy,

$$\frac{dU}{dt} = -kGM^2\left(\frac{2}{R^2}\right)_{R_1}^{R_0}\frac{dR}{dt} \approx L, \tag{2.2}$$

where the term dR/dt accounts for the required rate of change in the radius of the star to account for its observed luminosity, L. If we substitute for present day Sun values, taking $R_1 = R_{\odot}$ and $L = L_{\odot}$, and assume that R_0 is very large, so that the term $2/R_0^2$ in Equation (2.2) is effectively zero, the rate of change in the radius is $dR/dt = -7$

[4] If the matter is distributed uniformly, then $k = 3/5$.

meters per century. Such a small change in the Sun's radius, even over many millennia, would be completely unmeasurable from Earth, and it provides for what is called the Kelvin–Helmholtz (KH) or thermal timescale τ_{KH}. Given the KH timescale is effectively the time required for the Sun to radiate all of its internal energy into space, we have that

$$\tau_{KH} \approx \frac{U}{L} = \frac{GM^2}{RL} \tag{2.3}$$

In the case of the Sun, the KH timescale is of order 15 million years. We now have two evolutionary timescales for a star — the dynamic and the thermal. These timescales effectively limit the speed with which a star can react to some specific change. In the dynamical case, it is the collapse time for a star in which the thermal and material body have effectively separated from each other, and gravity is unopposed, effectively enabling it to eventually crush the star's material out of existence. In the thermal case, it is the time required for a star to radiate all of its internal thermal energy into space. Invariably, $\tau_{dyn} << \tau_{KH}$. In the case of the Sun, $\tau_{dyn} \approx 54$ minutes (taking the Sun's bulk density to be 1408 kg/m^3), and as seen earlier, $\tau_{KH} = 15 \times 10^6$ years. It turns out, however, these two timescales are dwarfed if a star can find an internal energy source to replenish the radiation that it loses into space at its surface. That the Sun must have found some internal energy source to maintain its luminosity is actually predicted by our understanding of geology on Earth. Given that the geologists place the age of the Earth well in excess of 15 million years (the Sun's KH timescale), the Sun must have tapped some additional internal energy source in order to account for its deduced age of some 4.5 billion years[5]. This internal energy source is, it turns out, derived from its own bulk. The Sun is effectively eating itself to keep shining. The crisis of stellar age and energy source came to a head in the early years of the 20th century, and it was Arthur Eddington who reasoned

[5] This is actually based upon the deduced ages of meteorites. The meteorites being derived from asteroids, which were the first-formed rocky objects within the solar system.

his way to the most likely energy source being tapped by the Sun and stars in general[6].

In terms of their physical structure, the two most common elements in the Universe, hydrogen and helium have one and two protons within their respective nuclei. On this basis, one might naively imagine that a helium atom might be assembled through the chance bringing together of two hydrogen atoms. Unfortunately, nature will not allow such a summation, and it transpires that it takes the interaction of six hydrogen atoms to make a helium atom. Even this latter statement is strange, however, since the atomic mass of helium is actually four times that of hydrogen. Considering such issues, Eddington was to write,

> "We used to think that if we knew one, we also knew two, because one and one make two. We are finding, however, that we must learn a great deal more about 'and'."

From a quantum mechanical perspective, ^{2}He — a two-proton nucleus — is unstable, and accordingly, as soon as it tries to form, it breaks apart again. In order, therefore, to make a helium nucleus out of hydrogen nuclei, an extra special set of circumstances needs to come about — this is the 'and' term that Eddington worried about in the early 1920s. Indeed, to get the ball rolling, what is needed is the coming together of two protons in such a way that one of them undergoes a transmutation to become a neutron (Figure 2.2). A neutron has the same mass as a proton, but as its name suggests, it carries no electric charge. Accordingly, if at the moment of closest approach of two protons one of them undergoes what is called an

[6] In his review of Chandrasekar's book, *An Introduction to the Study of Stellar Structure*, in the 22 July, 1939 issue of *Nature*, McCrea notes that one of the most striking features of the astrophysical theory of stars, up to that point in time, "is that so much can be done without assuming any particular mechanism of stellar energy generation". Indeed, a clear picture as to how stars might actually generate energy through fusion reactions, transmuting hydrogen into helium, was not apparent until the early to mid-1940s, following the pioneering work by Hans Bethe and Carl Friedrich von Weizsäcker in the late 1930s.

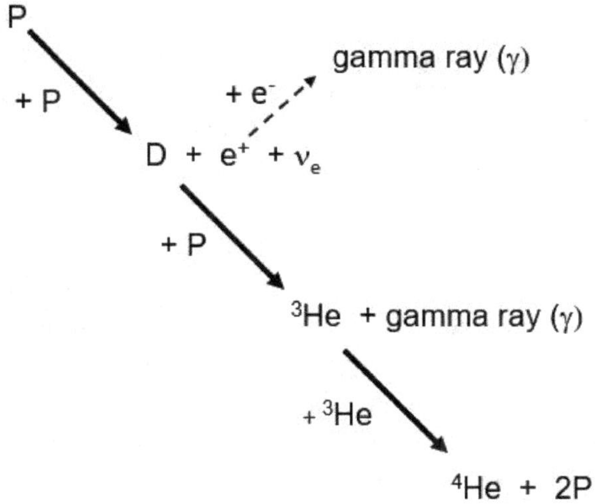

Figure 2.2. Schematic flow chart for the proton–proton chain by which a helium nucleus is 'assembled' from the interaction of six hydrogen nuclei. The process begins by the interaction of two protons enabling the formation of a deuterium, D, atom and the emission of an electron neutrino, ν_e, and a positron, e^+. The neutrino exits the star within a few seconds, while the positron will soon annihilate with an electron to produce energy within the star's interior. The deuterium atom quickly captures another proton to form a helium-3 nucleus, emitting a gamma ray photon, γ, in the process. Finally, two helium-3 atoms interact to produce a helium-4 atom, returning two hydrogen atoms into the star's interior at the same time.

inverse-beta decay to become a neutron, then a deuterium atom can form. Importantly, deuterium is stable and its nucleus consists of a proton and a neutron. What happens next is the coming together of a deuterium atom and a hydrogen atom to make helium-3 — that is an atom with a nucleus composed of two protons and a neutron. Importantly, again, this atom is stable. So far, three protons have been involved in the formation of ^{3}He, and we are just one step away from making helium-4. The final step requires two helium-3 nuclei to combine, and when this happens, we end up with our sought after ^{4}He. Since each ^{3}He atom is built-up from the interaction of three hydrogen atoms, six hydrogen atoms have been involved in the production of one ^{4}He atom. This is not the final story, however, since when two ^{3}He atoms interact, in addition to producing ^{4}He, two hydrogen atoms are returned to the star's interior. Accordingly, a total of four hydrogen atoms, from within the star, have been 'used-up' or 'corralled' to make one helium nucleus. The full chain of reactions, the so-called proton–proton or PP, chain is shown graphically in Figure 2.2.

Working out the physical details of the proton–proton chain, whereby four hydrogen atoms are eventually converted into a helium atom, is one thing. The reason why the chain is important to a star is another. Indeed, the conversion enables a star to generate energy within its interior, and it is this internal energy source that will enable a star to remain stable against gravitational collapse and continuously radiate energy into space at its surface. The energy tapped by the star comes from the annihilation of the positron, e^+ — formed in the formation of deuterium —, and in the gamma ray photon emitted during the interaction of the two ^{3}He atoms. Einstein's famous formula gives the positron annihilation energy as $2m_ec^2$, where m_e is the mass of the electron (and the positron — hence the factor of 2), and c is the speed of light. Through a myriad of interactions with the ions inside of a star's interior, the gamma ray photon will add to their thermal motion before it eventually exits the star, as a much lower energy photon, once it finds its way from the star's center to its photosphere. In total, the energy generated each time a new helium atom is formed from the bringing together and interaction of four hydrogen atoms is $E_{PP} = 4.4 \times 10^{-12}$ Joules. Einstein's mass–energy formula now accounts for the difference in the mass, ΔM, of four hydrogen atoms and that of the helium atom: $\Delta M = E_{PP}/c^2 = 4M_H - M_{4He} = 4.9 \times 10^{-29}$ kg. With these numbers in place, we can determine that the number of $4H \Rightarrow {}^4He$ conversions required per second, $N_{PP,}$ in order to power the Sun is: $N_{PP} \approx L_\odot/E_{PP} = 10^{38}$ conversions per second. This is a tremendously large number of interactions, resulting in some $\Delta M\ N_{PP} \approx 5 \times 10^9$ kg of the Sun's bulk mass being converted into energy each and every second in order to account for the energy radiated into space at its surface. Since the Sun formed, some 4.56 billion years ago, it has converted the equivalent of 117 Earth-masses of its original bulk into energy. This mass conversion seems large by human standards, but corresponds to just 0.04% of the Sun's present mass. Indeed, since the Sun is composed of about 10^{57} hydrogen atoms, at any instant, a minuscular fraction (just 10^{-19}) of its hydrogen atoms are actually involved in generating energy.

Stars are massive partly because they are feeding upon themselves, converting internal mass into energy through fusion reactions in their deep, dense, ultra-hot interiors. What this tells us is that when undergoing fusion reactions and converting hydrogen into helium, a star can tap each hydrogen atom within its core for an energy equivalent of 0.007 M_H. The nuclear timescale τ_{nuc} for a star of mass M and luminosity L will accordingly be of order

$$\tau_{nuc} = 0.007 M_H \left(\frac{fM}{M_H} \right) \frac{c^2}{L} \tag{2.4}$$

where $f \approx 0.1$ is the mass fraction of the star that is ultimately converted into helium — this number being derived according to a study of detailed numerical models. For the Sun, the nuclear lifetime for the conversion of hydrogen into helium, the so-called main sequence lifetime, is $\tau_{ms} = \tau_{nuc} = 3.3 \times 10^{17}$ seconds, or about 10 billion years and this, as is required by the geological observations relating to Earth's age, is much larger than the value of τ_{KH} seen earlier. From this derivation, we deduce that the Sun is about middle-aged, with another 5.5 billion years lying ahead of it before the hydrogen supply within its inner core begins to run out (that is, is converted into helium). In Chapter 1 it was indicated that main-sequence stars, i.e., those stars in the process of converting hydrogen into helium via fusion reactions in their cores, obey a well-defined mass–luminosity relationship of the form $L \sim M^{\alpha}$, with $1 < \alpha < 5$ depending on the characteristic mass range (recall Figure 1.13). Typically, for the main sequence, $\alpha = 3.5$ is used. With this relation, the main sequence lifetime will characteristically vary as $\tau_{nuc} = 10^{10}/M^{2.5}$ (years), where the mass M is expressed in solar mass units. This last result reveals an important point about star evolution: the more massive a star is, the more rapidly it moves through its hydrogen (that is, main sequence) phase. A 10 $M_\odot$ mass star will exhaust its core hydrogen supply in about 32 million years, moving some 316 times faster than the Sun through its main sequence phase. In contrast, low mass stars evolve much more slowly. A star with a mass just ¼ that of the Sun will move some 128 times more slowly than the Sun through its main sequence phase (that is, $\tau_{nuc} \sim 10^{12}$ years). Massive stars sprint through their main sequence phase, while low mass stars dawdle.

At the end of its main sequence phase, with the hydrogen supply in the central core approaching zero, a star needs to find a new energy generation source. As indicated in Figure 2.1, this new energy source is found within a star's helium supply. Energy is now generated by the conversion of three helium atoms into a carbon atom. Once again, the slight mass difference between the final carbon atom and the three generating helium atoms is accounted for through the production of energy — energy that the star can tap so as to keep its interior hot, and remain stable against gravitational collapse, as it continuously radiates energy into space at its surface. The conditions under which helium undergoes fusion reactions to produce carbon (the so-called triple-alpha reaction) are much more extreme than those required to convert hydrogen into helium. During its main sequence phase, when converting hydrogen into helium, the conditions inside the Sun's core have a characteristic density of 10^5 kg/m^3, and a characteristic temperature of 16 million degrees Kelvin. During its helium fusion phase, the Sun's core will have a characterised density of some 10^7 kg/m^3 and a temperature in the order of 200 million Kelvin.

In changing from one energy source to another, a star has to literally transform itself inside and out. To achieve the high density and temperature required for helium fusion to begin, the core of a star must shrink, becoming both smaller and hotter. In tandem with the core shrinking, the outer regions expand outward, becoming cooler and less dense. The star begins to become a red giant. Here the term red follows from Wien's law (recall Chapter 1.5), with the low temperature of such objects (just a few thousand Kelvin) specifying a peak energy flux in the red part of the electromagnetic spectrum. The giant part of the label relates to the size increase, which can be quite dramatic. In the case of the Sun, the radius will expand (at least temporarily) by a factor of several hundreds during the helium fusion stage. Once helium fusion begins, the Sun will have a characteristic surface temperature about ½ that at the present time, with a size some 10 times larger and a luminosity 40 times greater than those at present.

As shown in Figure 2.1, all stars will go through a hydrogen (main sequence) as well as a helium burning phase. What happens once the helium fuel supply is exhausted, however, depends critically upon the star's initial mass. Stars initially eight times more massive than the Sun can evolve into successive states that enable fusion reactions beyond that of helium to take place. The process is one of rapidly diminishing returns, however, and once the core is composed predominantly of iron, it runs into an energy generation crisis. Indeed, beyond the formation of iron, all further fusion reactions require the input of energy. Accordingly, the star eventually loses its fusion-based energy-generating capability, and the end phase that follows is rapid and catastrophic (see below). Below an initial mass of about eight times that of the Sun, stars are not able to rearrange their interiors into a state capable of supporting fusion reactions beyond that of converting helium into carbon. The carbon/oxygen rich central cores of such stars, however, evolve into a stable state supported by the degeneracy of electrons (see below). The white dwarf end-phase of these low mass stars is stable, and in principle, such remnants will remain stable for the rest of eternity[7].

2.4. And so, to rest

We need not hurry ourselves, Watson. We are safe from interruption.
(*His Last Bow*)

Stars eventually evolve into one of three final end states: a white dwarf, a neutron star, or a black hole. The end state is, once again,

[7]There is no eternal rest, of course, and eventually any given white dwarf (neutron star, or rogue planet) will be destroyed through a capture and accretion encounter with a black hole. This black hole-dominated universe, however, is set in the far, far, far distant future [3].

determined by the star's initial mass. Stars less than about eight times the mass of Sun will evolve into a white dwarf configuration. Stars more massive than this limit will become a neutron star if the initial mass is less than about fifty times that of the Sun, or a black hole if the initial mass is greater than about fifty times the mass of the Sun — the separation boundary between these latter two outcomes is still far from clear. White dwarfs and neutron stars will form when the effects of quantum degeneracy dominate, while a black hole will form if gravity dominates the final stages of evolution.

At play in determining the stability (and persistent longevity) of white dwarf and neutron star structures is the so-called Pauli Exclusion Principle (PEP). This is a quantum effect that places a limit upon the amount of 'crowding' that electrons, as well as neutrons, can endure. This resistance to overcrowding results in a temperature-independent degeneracy effect that can provide pressure support against gravitational collapse. When the conditions allow for electron degeneracy to operate, a stable white dwarf is formed. When the more extreme conditions of neutron degeneracy operate, then a stable neutron star is formed. Since degeneracy, once established, holds irrespective of temperature, a white dwarf or neutron star, once formed, will remain stable for all time, even as they cool off to temperatures close to absolute zero[8]. The very high-density conditions required for degeneracy dictate that white dwarfs and neutron stars are small, highly compact objects, and existentially constrained by an upper mass limit. A white dwarf with a mass equal to that of the Sun will have a size comparable to that of the Earth (about 12,000 km across). A neutron star with a mass similar to that of the Sun will be about 20 km across.

Neutron stars and white dwarfs represent the ultimate end phases in the life of low- and intermediate-mass stars, and unless influenced by some outside perturbation, these objects simply cool down, over billions of years, to a moribund and inactive state — they

[8]The ultimate temperature would be that corresponding to the background temperature of the universe. At the present epoch, this lower limit is set by the 2.7 K microwave background.

are, in some sense, the 'solid' stars discussed at the start of Section 2.3. While white dwarfs and neutron stars are stable for all time, once formed, there is a limit to how massive they can be. Subrahmanyan Chandrasekhar famously showed in the mid-1930s that white dwarfs are no longer stable if their mass exceeds a critical value of $M_{CWD} \sim 1.4\ M_{\odot}$. If the mass of a white dwarf, through binary companion accretion or the collision with another star, becomes larger than M_{CWD}, then the electron degeneracy condition can no longer hold sway, and a catastrophic collapse and conflagration must take place. Likewise, neutron degeneracy has its limits too, and once the mass of a neutron star exceeds $M_{CNS} \sim 2\ M_{\odot}$, catastrophic collapse is inevitable.

Beyond the mass limit set for the formation of a stable neutron star, gravitational collapse to form a black hole is inevitable. These latter objects can have any mass, but the extreme gravitational compression that they experience results in the formation of a singularity in which matter is crushed out of all existence. Indeed, our understanding of physics breaks down with the formation of a singularity, but fortunately, nature hides such non-reality states from our eyes behind the forever closed doors of an event horizon. The radius of the event horizon is effectively the distance from a singularity at which the escape velocity equals the speed of light. A black hole with the mass of the Sun will have an event horizon radius of about 3 km.

Acting as the ultimate trapdoor, anything that crosses over the event horizon boundary can no longer communicate with the outside world, and this dictates that we can never see what is actually taking place inside of a black hole — short of going there, but then we would not be able to tell anyone beyond the event horizon of the wonders that we had seen[9]. Black holes are the ultimate Vegas, and

[9] Black holes are both simple and remarkably complicated. They are simple in the sense that their entire structure is described according to their mass, spin and electric charge. They are complicated, however, because of the effects relating to the extreme gravitational forces at play near the singularity. For all that, the physics that we do know something about will take us well inside of the event horizon and most of the way to the central singularity.

they will ultimately inherit the entire Universe. Indeed, as the Universe ages, all matter (including white dwarfs and neutron stars) are destined, in the ultra deep, deep future, to end up being consumed by a black hole. The future of the Universe will eventually evolve into that of a starless, totally dark, black hole-filled domain [3]. Remarkably, however, it is generally thought that not even black holes last forever. So-called Hawking radiation, which originates at the event horizon boundary, results in the very, very, very slow evaporation of even the most massive black holes, and this suggests a final end state for the Universe as a vast expanse of spacetime containing nothing but radiation[10].

2.5. Sliming down

> [Introducing Mycroft Holmes] *His body was absolutely corpulent*
> (*The Adventure of the Greek Interpreter*)

At some level, all stars lose material from their outer layers into space. The Sun supports a wind that pushes a steady outward flow of charged particles into interplanetary space. The amount of material being lost by the Sun through this solar wind at the present time is minuscule compared to its total mass, amounting to about 10^{-14} solar masses per year (or some 317 million kg/s). Likewise, low mass dwarf stars lose mass from their surface regions through the action of magnetic field generation and the production of star-flares. The more massive a star is, however, the greater the amount of material that is lost from its surface. The most massive O and B spectral type stars can have mass loss rates on the main sequence as high as 10^{-5} to 10^{-6} $M_{\odot}$/yr, and these rates are sufficiently high to dramatically alter the course of their evolution.

In addition to the steady mass winnowing by stellar winds, stars can also slim their mass down in other ways. These include mass loss due to having a close stellar companion (discussed in Chapter 3),

[10] Here it is assumed that even electrons and protons ultimately decay — a condition that currently has no observational conformation or experimental support.

through the development of a large-scale instability in its outer layers, or through the catastrophic destruction of its central iron core. The former of these instabilities results in the formation of a planetary nebula, while the latter collapse results in the formation of a supernova.

The planetary nebula (PN) phase is a relatively gentle one, and proceeds over several tens of thousands of years. Stars less massive than about 8 $M_{\odot}$ undergo this phase upon the cessation of helium burning within their central cores. At this late stage in its evolution, the star is a highly luminous, large, bloated giant, with a relatively strong stellar wind developing at its surface. Indeed, the star can lose a significant amount of its initial mass at this stage, forming around itself an extended low density circumstellar envelope. In the case of the Sun, its final white dwarf configuration will have a mass that is about 60% of its present value. With the eventual emergence of the high-temperature white dwarf core, however, the material that has accumulated within the surrounding circumstellar envelope will become highly ionized[11], with recombination emission lines then being responsible for producing the visually spectacular planetary nebula display.

In contrast to the final PN stage of those stars having an initial mass below 8 $M_{\odot}$, more massive stars will undergo a dramatic and rapid end phase. These stars will literally collapse upon themselves, resulting in the short-lived appearance of a supernova. The key mechanism is the disruptive outward propagation of a shock wave, produced when a star's iron-rich core undergoes catastrophic disruption. Indeed, the shockwave literally rips the star apart, pushing material into the interstellar medium with speeds up to thousands of kilometers per second. Supernovae (SN) are literally the biggest bangs in the Universe since the formative Big Bang itself. Indeed, not only do supernovae return highly processed material into the interstellar medium, they also compress interstellar clouds, leading to the formation of new stars. From the accumulated ashes of SN, destroyed

[11] Since newly formed white dwarfs have very high surface temperatures, Wien's law dictates that they will radiate copious amounts of ionizing, ultraviolet radiation.

Table 2.1. A brief synopsis of supernova types. The first and second columns indicate the SN type and the corresponding progenitor body (whether it is a star, white dwarf or neutron star). The third column indicates the key physical process that triggers the SN to occur, while the fourth and fifth columns indicate the key observational properties of the SN disruption and the characteristics of the remnant object.

	Progenitor	Process	Spectrum/ observation	End result
Type II	Single massive star ($M > 8\ M_\odot$)	Collapse of iron core	Hydrogen present	Neutron star
Type Ia	White dwarf	Deflagration	No hydrogen present	disruption
Kilonova	Neutron star	Merger	Gamma-ray burst	disruption
Hypernova	Single massive star ($M > 30\ M_\odot$)	Collapse of iron core	Gamma-ray burst	Neutron star/ Black hole

stars rise, phoenix-like, the flickering lights of a new generation of stellar bodies.

Many different forms of supernovae are now recognized, with the various types being distinguished according to their spectrum characteristics, light curve morphology, location and environment, as well as progenitor body. Table 2.1 provides a highly simplified overview to the main SN types and their formation channels. Type II supernova are produced when the iron-rich cores, within stars initially more massive than about 8 $M_\odot$, undergo catastrophic collapse. These SN are only observed in spiral galaxies, and specifically in their spiral arm features. The optical spectra of such SN typically reveal hydrogen lines, and the remnant object, if there is one, is a neutron star. Type Ia SN are produced when a white dwarf accretes sufficient material (from a binary companion — see Chapter 3) to push it over the Chandrasekar stability limit. The process proceeds when the degeneracy is lifted in its outer layers, enabling an aggressive, runaway process involving carbon fusion reactions (so-called carbon deflagration) to come about. The end result is the complete disruption of the white dwarf. As Type Ia SN are associated with a breach of the Chandrasekar limit, they are assumed to be 'standard bombs', for which the peak brightness should be the same for all such events. In this manner, Type Ia SN are extremely important in determining the distances to the most remote of galaxies. Kilonova are taken to be the result of two neutron stars merging, with the merger resulting in the formation of an object with a mass greater than the stability limit for degenerate neutron matter. These SN produce energetic gamma-ray bursts, and seemingly leave no remnant object. Hypernova are thought to be produced via the catastrophic collapse of an iron-rich core within a massive star, but now the initial mass is greater than some 30 $M_\odot$. Hypernova SN produce strong gamma-ray bursts, with the remnant being either a neutron star or a black hole.

2.5.1. *Coda: Apple pies, and the making of us*

> *"Elementary!" Said he.*
>
> (*The Adventure of the Crooked Man*)

Astronomer Carl Sagan once remarked that to make an apple pie from scratch, you must first create a Universe. In this statement he is entirely right, and central to being able to make that apple pie is the formation of stars, since stars are the chemical reactors of creation. The Big Bang start to the Universe generated the hydrogen and helium that is the raw material for making stars, and the stars, through the various fusion reactions operating within the interiors, together with the massive stars, through their final supernova disruption phase, generating all the other stable elements in the periodic table. All the carbon, oxygen, nitrogen, sodium and iron in the food we eat, and indeed within our very makeup, were derived from countless earlier generations of stars. The Earth, us, the oceans, the forests, the atmosphere, and everything around us is composed of atoms and compounds produced by stars acting upon the initial cosmological hydrogen and helium. The stars, by being element producing factories, have changed, and will continue to change for many trillions of years yet, the very makeup and destiny of the Universe, taking its initial bountiful but sterile store of hydrogen and helium, and turning it into fecund matter capable of generating wonderful entities and incredible structures.

That the Earth, us, and apple pies exist is the result of stars forming and evolving. Indeed, the vast reservoir of elementary atoms within the interstellar medium, is slowly, but assuredly, being turned into new stars, and as each new generation of stars evolve, nuclear processed material is recycled back into interstellar space. In this manner, the composition of the interstellar medium is gradually changing, containing more and more metals (elements, recall, other than hydrogen and helium) as time ticks forward. The recycling is not 100% efficient, with some of the raw interstellar material being turned into white dwarfs, neutron stars, and black holes — objects that lock and trap material away from future recycling and processing. For all that, given enough time, the processed atomic dregs and dust of once-active stars build up to a level that when a new generation of stars form, they can begin to be accompanied by terrestrial planets — indeed, planets such as the Earth. To make an apple pie, you need to first evolve apple trees, on a nurturing terrestrial planet, orbiting at the just-right distance from a solar mass star, and before all this, you need to allow time for the formation and death of innumerable numbers of stars to generate the fundamental atoms for molecular constructions. It took the universe about 13.7 billion years to make the reality of apple pies possible.

2.6. Journeys through the HR diagram

Just look up the trains in Bradshaw
(*The Copper Beeches*. [*Bradshaw's* was a well known and popular series of train timetables and travel guides])

Having begun generating energy, via nuclear fusion, within its core, a star begins the process of transporting that energy outward — ultimately, at the surface (the photosphere), the energy generated in the core is radiated into the vast coldness of space. For this to happen, the temperature must decrease outwards from the core, with $\Delta T/\Delta R$

~ $(T_c - T_s)/R$ where the subscripts c and s indicate central and surface temperatures and where R is the radius of the star. In the case of the Sun, the central temperature is some 15 million Kelvin, and accordingly, $\Delta T/\Delta R$ ~ 0.02 K/m indicates that thermodynamic equilibrium holds to a good approximation at each location within the interior. Despite this latter statement, energy does (indeed, must) flow outward, down the temperature gradient. Indeed, it is the temperature gradient that allows for the establishment of a pressure gradient (with the pressure decreasing outward from the center) which supports the weight of overlying layers. In this manner, the onset of nuclear reactions enables a star to remain stable against gravitational collapse — accordingly, a dynamical balance is established, and this will hold as long as nuclear fusion reactions can keep generating energy. This latter condition, in turn, being determined by the fuel supply, which on the main sequence, is the store of hydrogen within the central regions. Once all the hydrogen within a stellar core has been converted into helium, however, a new fusion reaction must be found if the star is to regain a relatively stable state against gravitational collapse.

In broad brush terms, the main sequence luminosity, L, of a star (that is, the energy radiated into space at its surface) is related to the rate at which hydrogen is being used up in the core, dX/dt, where X is the hydrogen mass fraction per kilogram. The amount of energy generated per kilogram by fusion reactions per second is $Q = 0.007\ c^2$, and the total mass $f\ M$ of hydrogen within a star that can be tapped in order to fuel the fusion reactions in its central region (accordingly, we have $0 < f < 1$, with M being the total mass of the star). From the above, we can formally write an expression for a star's luminosity in terms of its mass and the rate of change in its hydrogen abundance [4]:

$$L = fQM\left(-\frac{dX}{dt}\right) \tag{2.5}$$

The size of the energy generating region within a star can be estimated from the average temperature gradient, $\Delta T/\Delta R$, and the temperature limit T_F for hydrogen fusion reactions to proceed. In this manner, for the Sun, taking T_c = 15 million Kelvin and T_F = 10 million Kelvin, we have

$$(T_c - T_F)/(\Delta T/\Delta R) \sim 5 \times 10^6/0.02 = 2.5 \times 10^8 \text{ meters} \sim R_\odot/3 \qquad (2.6)$$

That is, about 1/3 of the Sun's interior by radius is involved in generating energy through hydrogen fusion reactions. Due to the central mass concentration within the Sun, however, this radial extent encompasses a region containing about 60% of the Sun's total mass.

Since the demands upon the central energy generation process become increasingly severe as the mass of a star increases, the mode by which energy moves through the interior of the star changes. In lowest mass stars, $M < 0.3\ M_\odot$, energy transport is largely by convection, with the interiors being nearly fully mixed and chemically homogeneous. Between $0.3 < M/M_\odot < 1.5$, the transport of energy away from the central regions is largely through radiation rather than the bulk motion of the internal plasma. In the case of the Sun, only the outer 1/3 by radius is convective. Beyond about 1.5 $M_\odot$, energy transport in the central region becomes convective — this being because of the increasing importance of a second process by which hydrogen can be converted into helium[12]. These different energy transport mechanisms and domains have important consequences for the future evolution of a star as hydrogen is increasingly consumed and reduced in the central core. Importantly, however, detailed numerical calculations indicate that as the composition difference between the core and envelope of a star increases (through the action of fusion reactions within the core) the luminosity and radius of a star increase, while its surface temperature decreases. As it ages, a star moves to the right and upwards in the HR diagram — mapping out, as it goes, an evolutionary track.

[12]This is the CN cycle which uses carbon as a catalyst nucleus to enable to the conversion of 4 hydrogen nuclei (protons) into a helium nucleus.

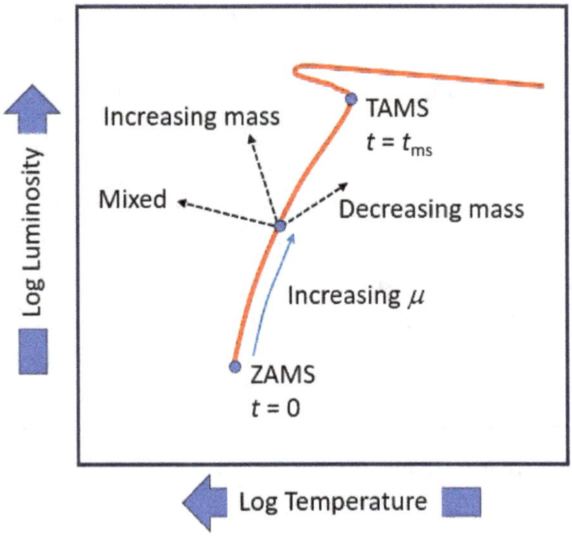

Figure 2.3. Schematic evolutionary path ways for a main sequence star (red line). Starting at the zero-age main sequence (ZAMS) location, corresponding to time t = 0, to the termination of the main sequence (TAMS) location, and the exhaustion of hydrogen within the star's central core, at time $t = t_{ms}$. In general, evolution is away from the ZAMS towards greater luminosity, larger radius, and lower surface temperature. By introducing mixing events, mass loss, or mass accretion (at some specific time: $0 < t < t_{ms}$), the evolutionary track can be significantly modified (dashed lines). See text for details.

Figure 2.3 shows a schematic evolutionary track, in an HR diagram, for an intermediate mass star from its initial zero-age main sequence (ZAMS) location, at time $t = 0$, through to its termination of the main sequence (TAMS) location, at which moment hydrogen is exhausted within its central core (at a time $t = \tau_{nuc} = t_{ms}$). The time is set to $t = 0$ on the ZAMS since prior to this time, although the star has been evolving from some proto-stellar state, and undergoing gravitational collapse, becoming smaller and hotter, it does this without changing its composition[13]. The ZAMS location corresponds to

[13] Another way of saying this is that a star has no memory of its assembly process and/or evolutionary times prior to the onset of hydrogen fusion reactions.

the star's luminosity, temperature, and radius at the moment core fusion reactions are initiated, with evolutionary changes, thereafter, being driven by changes in internal composition. Indeed, a star evolves, that is, its luminosity, radius, and surface temperature change over time because there is an increasing difference between the core and envelope compositions. This effect is parametrised according to the mean molecular weight, μ. Through this latter term, account is taken of the number density of electrons and ions $n(r)$ present, at a specific location r, within a star. This variation is important since the gas pressure P_{gas} at each location within a star's interior varies according to the relationship $P_{gas}(r) = n(r)\ k\ T(r)$, where k is the Boltzmann constant, and T is the temperature[14]. Each fully ionized hydrogen atom provides 1 electron and 1 proton to a star's plasma; helium provides 2 electrons and an α-particle composed of two protons and two neutrons; and so on for other elements. In terms of a star's hydrogen, helium, and heavy element mass fractions X, Y, and Z (recall Chapter 1.2), the mean molecular weight varies as: $1/\mu = 2X + 3Y/4 + Z/2$, where μ is defined according to the relation: $n(r) = \rho(r) / (\mu\, m_H)$, where $\rho(r)$ is the density at radius r, and m_H is the mass of the hydrogen atom [see e.g., reference 3]. A star composed entirely of hydrogen (X = 1.0) will have a mean molecular weight of μ = ½. A pure helium star (Y = 1.0) will have a mean molecular weight of $\mu = 4/3$. The Sun, with its original ZAMS compositional mass fractions of X = 0.71, Y = 0.27, and Z = 0.02, has a mean molecular weight of $\mu = 0.613$. For a star made of just hydrogen and helium (that is with Z = 0.0), the mean molecular weight will vary as $\mu = 4 / (3 + 5\ X(t))$, where now we are showing the specific time variability of the hydrogen mass fraction term. This term varies, recall, as a result of the fusion reactions taking place within a star's core, and the steady conversion of 4 hydrogen atoms (that is, 4 protons) into an α-particle, that is, the nucleus of a helium atom (recall Figure 2.2). The mean molecular weight in the outer layers of a star,

[14] There is also a radiation pressure term: $P_{rad} = a\ T^4 / 3$, where a is the radiation density constant, and T is the temperature. The total pressure at each point in the interior is given by $P = P_{rad} + P_{gas}$.

situated well away from the region of fusion reactions, remains constant. For our hydrogen–helium star, with, say, $X = 0.7$ and $Y = 0.3$, the variation in the core mean molecular weight will vary from μ(ZAMS) = 0.615, to μ(TAMS) = 1.333. Where all this comes into perspective is that the luminosity of a star, with a mass something like that of the Sun, during its main sequence phase, changes as the mean molecular weight to the power 7.5 [see e.g., reference 4]. Accordingly, between the ZAMS and the TAMS, the luminosity of a star will increase by a factor of something like $(1.333 / 0.615)^{7.5} \approx 15$. Now many other important things are going on during a star's main sequence evolution, but to first order we have the result that, by simply allowing fusion reactions to convert hydrogen into helium, the luminosity of a star should increase by about an order of magnitude (factor of about 10) between its ZAMS and TAMS locations. That a star should increase in size and decrease in its surface temperature as it evolves through its main sequence phase is less easily explained [see e.g., 4], but the luminosity, temperature and radius will, at all times, satisfy the Stefan–Boltzmann condition given in Equation (1.6).

The vocabulary of Bradshaw is nervous and terse, but limited
(*The Valley of Fear*)

Anticipating future discussion with respect to the formation of blue stragglers, Figure 2.3 illustrates the results of 'tinkering' with a model star as it evolves. If at some particular time *t*, mass is added to a star (perhaps through a binary companion or through a collision), the evolutionary path will be 'diverted' towards attaining a higher luminosity (a result, partially, of the mass–luminosity relationship — recall Figure 1.12) and a higher temperature than those indicated by the canonical track (the red track in Figure 2.3). Likewise, if mass is lost (through a grazing close encounter with another star perhaps), then the luminosity will track at a lower value than that shown by the canonical track. If the mass remains constant, but the interior becomes fully-mixed, then the evolution will tend towards both higher luminosity and higher surface temperatures.

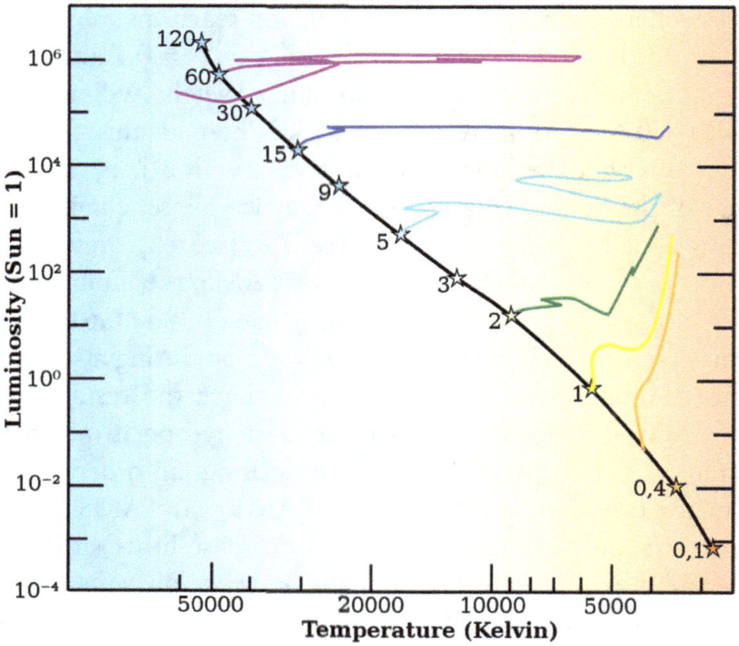

Figure 2.4. The evolutionary tracks for various mass stars in the HR diagram. The diagonal running from the upper left to the lower right (black line) corresponds to the zero-age main sequence, and is defined according to the luminosity, temperature, and radius of a star (of given mass) at the onset moment of hydrogen fusion reactions. Image from https:// www.collegesidekick.com/study-guides/earthscience/stellar-life-cycle.

The evolutionary tracks to a series of stars with masses between 0.1 and 60 $M_{\odot}$ are shown in Figure 2.4. The solid black line in the diagram indicates the zero-age main sequence, and indicates the luminosity, temperature and radius at the moment that fusion reactions first begin to run within a star's interior. The time each star spends in the main sequence phase, that is, the time over which it generates energy by hydrogen fusion reactions, is determined by the nuclear timescale τ_{nuc} as given earlier in Equation (2.4). With the exhaustion of hydrogen within its central core, a star moves rapidly (on a thermal τ_{KH} timescale) to the right in the HR diagram, becoming a lower temperature, higher-luminosity, larger-sized, giant.

Eventually, the interior of a hydrogen-depleted–core star restructures itself so that helium fusion reactions through the triple-alpha process can occur, with (schematically) 3 He ⇒ C + energy. The manner in which a star evolves across the HR diagram, from its zero-age main sequence location, varies according to mass, with various loops occurring in the tracks of the more massive stars. These loops occur with respect to conditions relating to mass loss and the onset of so-called shell burning phases, where fusion reactions occur in a spherical shell about an otherwise inert central core. While massive stars evolve across the HR diagram at an essentially constant luminosity, low mass stars show a significant increase in energy output prior to the onset of helium fusion reactions. At the tip of the red giant branch, the Sun, for example, will briefly attain a luminosity in the order of 1,000 times greater than its present vale. Once core helium fusion reactions have begun in the Sun's core it will settle down to a luminosity about 50 times greater than at present.

The evolutionary tracks shown in Figure 2.4 cover very different timescales. Indeed, as seen earlier, the more massive a star is, the more rapidly it evolves through its various energy-generation phases to a final end state. Figure 2.5, in contrast, shows a series of, so-called, isochrones (equal evolutionary time loci) which link together the luminosity and temperature locations of stars of different mass but with the same age. Such isochrones are particularly useful for estimating the ages of star clusters, this being predicated on the idea that all the stars within the cluster formed at the same time. The top isochrone (solid black line) in Figure 2.5 corresponds to different mass star locations in the HR diagram when the common age is 30 million years. At this age, stars in the mass range of 8 to 9 $M_\odot$ will have left the main sequence and will be in the process of initiating helium fusion reactions within their cores (as red giants). Stars with mass of about 7 $M_\odot$ will be closing in on their terminal (TAMS) main sequence phase. At 100 million years of age, stars a little more massive than 4 $M_\odot$ will have evolved away from the main sequence, to become giants, while stars less massive than about 3 $M_\odot$ will have barely moved away from their zero-age main sequence locations. At

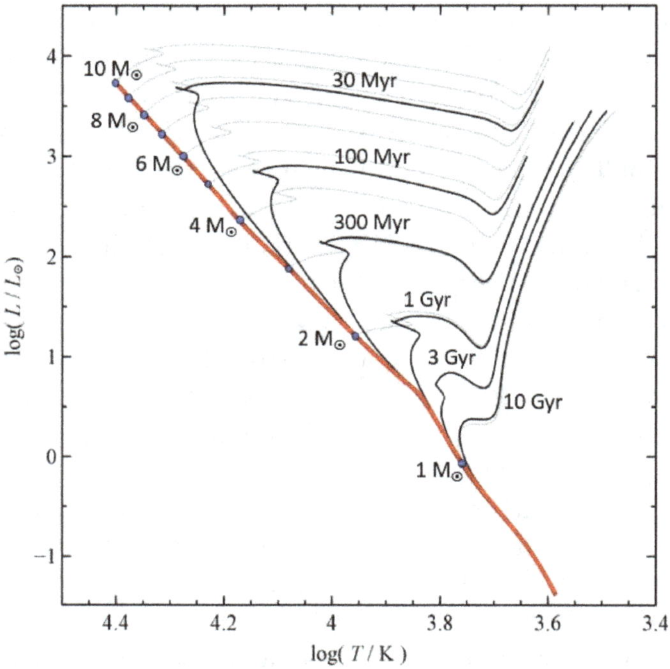

Figure 2.5. Isochrones (solid black lines) constructed from the evolutionary tracks (light grey lines) of stars having masses between 1 and 10 $M_\odot$. The red line corresponds to the zero-age main sequence, and the isochrones are interpolated according to the temperature and luminosity appropriate to each star at ages corresponding to 30 million, 100 million, and 300 million years, as well as 1 billion, 3 billion, and 10 billion years. The more massive stars move, or peel away from the zero-age main sequence sooner than lower mass ones. Image adapted from https://physics.stackexchange.com/questions/411298/stellar-isochrones-what-are-they

300 million years of age, all those stars with initial masses greater than 3 $M_\odot$ will have evolved away from the main sequence, while stars less massive than about 1.5 $M_\odot$ will still be located close to their zero-age main sequence positions. At 10 billion years of age, all stars more massive than about 1 $M_\odot$ will have evolved off the main sequence. By comparing the observationally deduced HR diagram for a cluster against the theoretical isochrone loci (i.e., as per Figure 2.5), the cluster age, along with the so-called turn-off mass, M_{TO}, can be determined. Stars more massive than M_{TO}, in a specific

cluster, will no longer be on the main sequence, but will have evolved into the giant region of the HR diagram, settled into some final white dwarf end state, or, for the most massive stars, undergone supernova disruption.

2.7. Cluster diagrams — the stragglers emerge

Things are moving very fast now and quite in accordance with the time-table

(*His Last Bow*)

An estimate for the turn-off mass, M_{TO}, in a cluster of age T can be found by rearranging Equation 2.4, and accordingly,

$$M_{TO} \approx \frac{2}{(T/1\,\mathrm{Gyr})^{0.4}} \tag{2.7}$$

where in (2.7) the age is cast in units of billions of years. At 1 billion years of age, for example, all stars more massive than about 2 $M_\odot$ will have evolved from the main sequence towards the giant branch, undergone supernova disruption (if more massive than 8 $M_\odot$), or become a white dwarf (if less massive than 8 $M_\odot$). At 30 million years of age, all stars more massive than about 8 $M_\odot$ will have evolved away from the main sequence.

Figure 2.6 shows the observational data, with respect to a color–magnitude diagram, for the globular cluster M 3. This particular globular is located about 10 kpc from the Solar System, contains about half a million stars, and, from the estimated turn-off (TO) position (at V = 19.2, B-V = 0.3) it is estimated to be about 11.4 billion years old. With this age, the turn-off mass for M 3 is $M_{TO} \sim 0.8$ $M_\odot$. Stars slightly more massive than M_{TO} will be situated on the red giant branch (RGB) or the horizontal branch (HB) where they are in in the process of converting helium into carbon within their central cores, or on the asymptotic giant branch (AGB) generating energy by fusion reactions within hydrogen and helium shell-burning regions about the core. Stars more massive than about 1 $M_\odot$ in M 3 will have

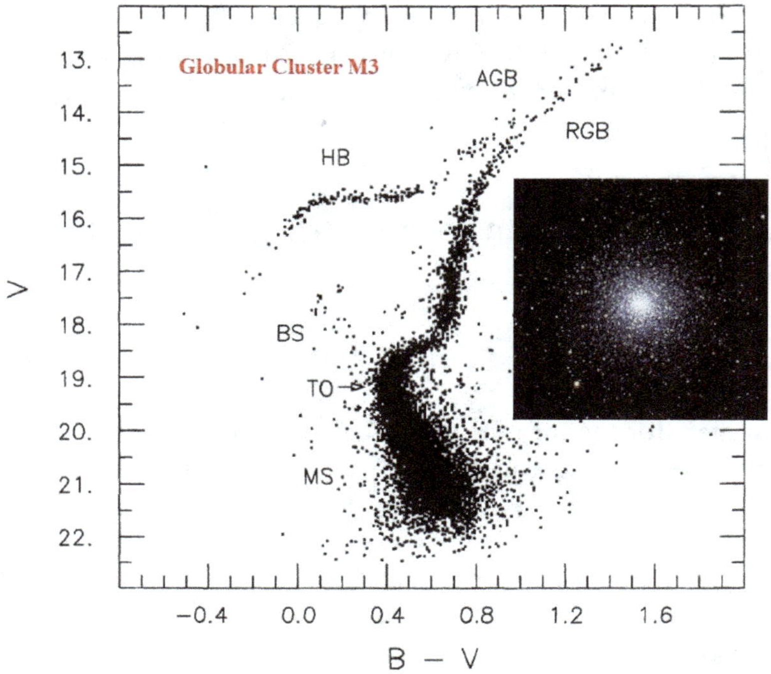

Figure 2.6. Color–magnitude diagram for globular cluster M 3. The various labels shown correspond to: MS = main sequence; TO = turn-off location; BS = blue stragglers; RGB = red giant branch; AGB = asymptotic giant branch; and HB = horizontal branch (see text for details). The inset on the middle-right shows the optical appearance of M 3 — image by Adam Block (Mount Lemon SkyCenter / University of Arizona).

run their entire evolutionary course to a final white dwarf phase, or supernova disruption if more massive than about 8 times the mass of the Sun (recall Figure 2.1). All this discussion provides for the canonical picture of stellar evolution, assuming that all the stars in the cluster are single or members of well separated binary systems. Furthermore, and importantly for our story, Figure 2.6 highlights a problem — a feature that has no clear standard stellar evolutionary explanation. Specifically, what is seen in M 3 is that a handful of stars are plotted in the color–magnitude diagram on the extended main sequence above the turn-off point derived for the cluster (that is, these stars have a higher luminosity and greater temperature than a star with a mass M_{TO}). The implication for these stars, indeed, these blue straggler (BS) stars, is that they have masses greater than M_{TO}, but for some reason have not evolved away from the main sequence. These stars are hotter (that is, they are bluer in color) and more luminous (have smaller visual magnitudes) than the temperature and luminosity set by the cluster turn-off mass, M_{TO}. Accordingly, we infer that $M_{BS} > M_{TO}$ — and yet, of course, this shouldn't be so.

2.7.1. *Are there really blue stragglers?*

> *Sometimes to solve a case, one must first solve another*
> (*Sherlock - TV series: The Abominable Bride*)

One possible first response to the appearance of the blue straggler (BS) region seen in Figure 2.6 is to deny its real existence. Not to say that the stars plotted there don't exist, but rather that they are 'normal' stars seen under unexpected circumstances. Indeed, scientists by nature are highly conservative, and suspicious of the new and the unexpected. This is a good trait, but it can often stifle progress. Nonetheless, it is entirely reasonable to consider ways in which the apparent problem of blue stragglers can be done away with. We have seen that the blue stragglers stand out in the colour–magnitude diagram because they plot in a region that, under standard stellar evolutionary theory, should be devoid of stars. This observation certainly calls out for an explanation, but the explanation could be something entirely mundane. Perhaps, for example, it might be argued that we are actually seeing stars that do not belong to the cluster under observation, and that they are ordinary stars that chance to be in our line of sight to the cluster. Or, perhaps they are simply stars that have poorly observed fluxes — the result of some inexplicable observational glitch. Indeed, these arguments need to be thoroughly addressed, and dismissed, before we can be sure that we are actually seeing a distinct group of stars. Without dragging out the debate, however, astronomers have successfully addressed these issues, and the vast majority of blue stragglers observed in clusters are indeed cluster members, and they have been observed correctly.

Another possible way to dismiss the potential existence of blue stragglers is according to the argument that they are actually perfectly normal[15] stars, seen in a relatively rare, and hitherto unexpected phase of their evolution. Are the blue stragglers really familiar objects seen in an unfamiliar way? This, indeed, was the position in the late 1950s and the 1960s when it was still unclear exactly how stars evolved as they aged. Writing in the *Astrophysical Journal* in 1964, Olin J, Eggen and Alan Sandage (see the next section), for example suggested that, "these stars [the then putative blue stragglers] may be moving back to the main sequence from the tip of the giant branch." In this manner, the supposed blue stragglers were ordinary (as such) sub-dwarf stars undergoing helium fusion reactions within their central cores, and were simply populating the extreme blue (that is, higher temperature) tail and hook of the horizontal branch (HB in Figure 2.6). The so-called blue stragglers are therefore just evolved stars that chance to fall close to, and above, the main sequence turn-off in a cluster's HR diagram. By this argument, there are no blue stragglers; they are just extreme examples of blue horizontal branch stars. Indeed, detailed numerical modeling indicates that one way to change the position at which a star falls on the horizontal branch in the HR diagram is to vary the mass of its hydrogen envelope and/or change its metallicity (or helium content). The more massive the hydrogen envelope, the larger and cooler a horizontal branch star is; the higher the helium abundance (the lower the metallicity), the higher the luminosity and the hotter the star (for a given mass). In this manner, one might argue that the so-called blue stragglers are just an odd sub-group of extreme horizontal branch stars, explainable according to appropriate adjustments in the mass of their hydrogen envelopes and helium abundance. This latter condition is difficult to disprove, but if it is the answer to the problem, then one is stuck with a fine-tuning issue, and in need of an explanation of why some HB stars evolve differently to other cluster members with the same mass.

[15] 'Normal' here is somewhat loosely applied, but the point is that the star has not suffered from some external interaction, or formed under rare and/or unlikely conditions.

It is easy to find reasons not to believe a specific observational result, especially if that observational result is not expected, but after a while it becomes clear that continued denial is unjustified — the body of evidence making it clear that what is being seen is, indeed, something new. When it comes to the blue stragglers, we can now be sure that they are a distinct group of stars, and not merely familiar stars seen in unfamiliar, misidentified, or unexpected ways.

2.8. The stragglers observed

Facts are better than theories

(*The Sign of Four*)

It is fair to say that the discovery of blue stragglers, within the color–magnitude diagram of one globular cluster or another, was an inevitability, although no one was specifically looking for such a population of stars. In 1953, however, the observer who dared to observe rather than simply see, was a young doctoral student at the California Institute of Technology, Allan Sandage[16]. As part of his doctoral work, Sandage was tasked with constructing the color–magnitude diagram for M 3 (our Figure 2.6). Using the 100- and 200-inch telescopes at Mount Palomar, photographic data on more than 1,100 cluster stars was collected. With this data, Sandage deduced from the cluster turn-off location that the cluster was some five billion years old (about ½ the accepted age value in the modern day). Furthermore, Sandage compared the color–magnitude diagram of M 3 with that previously constructed (with Halton Arp and William Alvin Baum) for the globular cluster M 92. From this comparison, Sandage noted that the cluster diagrams differed in that M 3 showed a distinct set of stars above the cluster turn-off (these are the blue stragglers). There was no indication of such stars existing above the turn-off point in the color–magnitude diagram for M 92.

[16] Sandage is perhaps better known in the modern era for his pioneering work in the determination of the Hubble constant, which is a measure of the expansion rate of the universe.

At this stage, Sandage drew attention to the "anomalous blue stars" of M 3, but left them, as such, undefined. The name "blue straggler" first appeared in print in a publication by Margaret Burbidge and Allan Sandage in the September 1958 issue of the *Astrophysical Journal* [5]. In this particular publication, Burbidge and Sandage were studying the color–magnitude diagram of the open (galactic) cluster NGC 7789, and upon drawing attention to the existence of stars blueward of (that is, hotter than) the turnoff for the cluster, noted that they were particularly interesting because "these bright blue stragglers are not understood on current ideas of stellar evolution." It was also clear at this stage (the late-1950s) that blue stragglers were probably a real phenomenon since their presence had been identified in the color–magnitude diagrams of not only M 3 and NGC 7789, but also in those for M 67, the Praesepe cluster (M 44), and the Coma Berenices cluster (Collinder 256). For all that, the phenomenon of blue stragglers remained a theoretical mystery. In his Presidential Address to the assembled fellows of the Royal Astronomical Society, on 9th February, 1962, McCrea summed up the situation as follows,

> "… the cluster M 67 is recognized to be amongst the oldest galactic clusters and it does most things by the book[17]. But the observers find that it also does something that clusters of its age ought not to do. It contains some blue stars lying on the continuation of the main sequence beyond the turn-off point of the cluster sequence [6]."

One can have a recognized observational phenomenon, but until there is also a theoretical understanding of the phenomenon, there is, in effect, a mystery. That is, we do not understand what it is that we are seeing. The development of a theoretical foundation behind an

[17]By this, McCrea meant that the cluster's HR diagram is consistent with expectation, given that all of the stars formed at the same time and had the same initial composition. On this basis, blue stragglers do not comply with expectation, or, as McCrea puts it, "the book."

observational phenomenon is all about generating an understanding of what that phenomenon is telling us.

Fundamental to all science is the collection of data, and the fundamental question with respect to blue stragglers, is not just how they form, but where and how often. Data on these latter two issues can be gathered by counting — it's a simple as 1 + 1 + 1 and so on. Well, actually, its not quite that simple, but the idea is that one can count the apparent number of blue stragglers in a given cluster, and then compare that count to other quantified properties of the cluster, such as its age and total mass. The term "apparent" is used with respect to the numbers since there are always going to be some that are missed through being too faint, or simply lost in the crowded field of other stars. In spite of this, Figure 2.7 shows a plot of the number of blue stragglers observed in a sample of open clusters against cluster age. Here, an immediate result jumps out to the eye, and this is that clusters much younger than about 100 million years in age tend to have very few blue stragglers — not necessarily zero, but very few. Clusters older than about a billion years, in contrast, have increasing numbers of blue stragglers. This 'visual' result is highly suggestive; it argues for

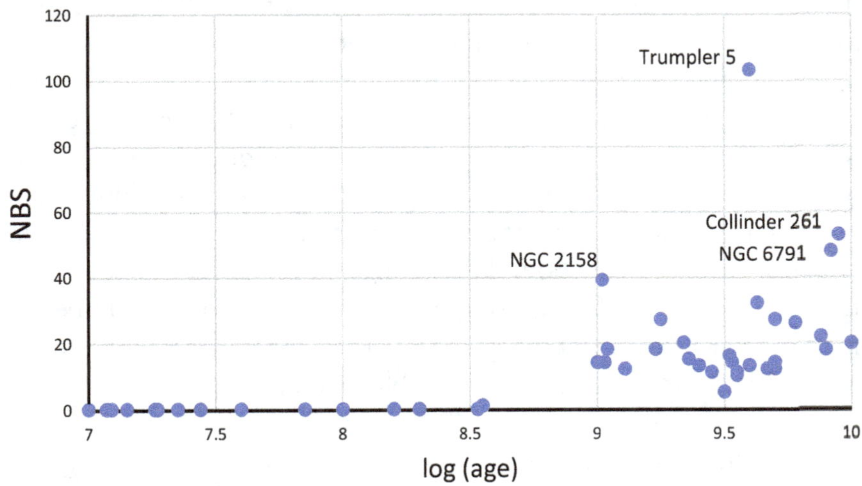

Figure 2.7. Number of blue stragglers observed within a cluster versus cluster age. A selection of cluster names is provided for reference in the text. Data from [7].

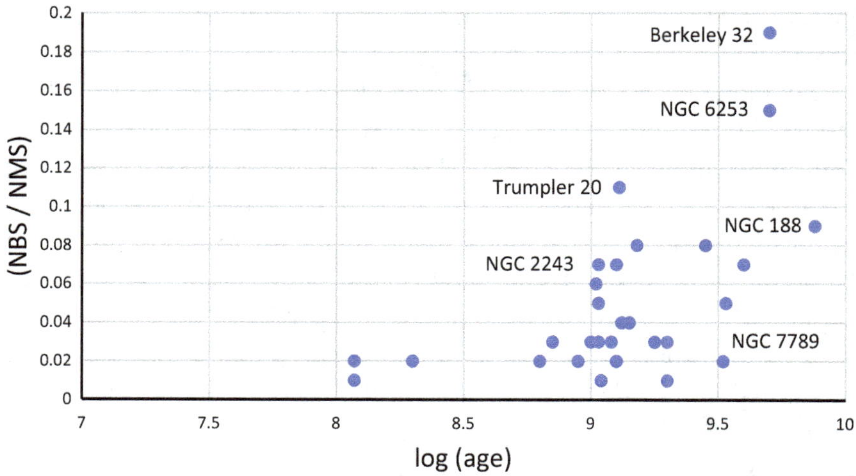

Figure 2.8. Number of blue stragglers relative to the number of main sequence stars, plotted against cluster age. A selection of cluster names is provided for reference in the text. Data from [7].

the notion that whatever mechanisms are responsible for blue straggler formation typically take time to work, and work more efficiently in clusters older than about one billion years of age. While the total number of blue stragglers that might be observed in a given cluster is highly variable, and variable as a function of cluster age, at the present time, Trumpler 5 (also cataloged as Collinder 105) has the greatest pure number count, although it is not the oldest known cluster. Figure 2.8 shows a similar plot to Figure 2.7, but now the blue straggler count is presented as a ratio, with respect to the number of main sequence stars, as a function of cluster age. In this case, it appears that the relative number of blue stragglers normalized to the number of main sequence stars[18] increases with cluster age. Once again, we see an age effect coming into play, with blue stragglers being more common in older clusters.

In addition to looking for correlations between blue straggler numbers and cluster age, one can also look at other cluster

[18] This is actually the number of main sequence stars up to 1 magnitude below (that is, fainter than) the deduced cluster turn-off magnitude.

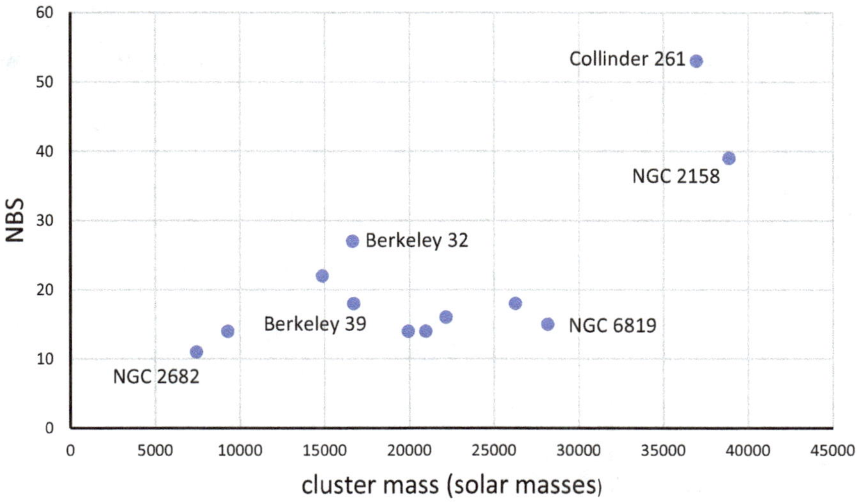

Figure 2.9. Number of blue stragglers within a cluster versus total cluster mass. A selection of cluster names is provided for reference in the text. Data from [7].

characteristics such as total mass. Figure 2.9 shows a plot of the number of blue stragglers versus total cluster mass (where this number is actually known). Once again, an interesting correlation appears to exist, with the number of blue stragglers increasing in lockstep with cluster mass. A detailed analysis of the data indicates that the number of blue stragglers in a cluster increases as the cluster mass taken to the one-half power: $N_{BS} \sim M_T^{0.5}$. The reasons for this (relatively) orderly relationship remains, at the present time, largely unexplained.

In addition to age, total cluster mass, and the relative number of blue stragglers to main sequence stars, another fundamental characteristic of any cluster is its binary star fraction f. This fraction is literally the ratio of the number of binary systems within the cluster compared to the total number of stars within the cluster. Accordingly, $0 < f < 1$, with $f \sim 0$ indicating essentially no binary systems within the cluster, and $f \sim 1$ indicating a cluster composed almost entirely of binary systems. The binary fraction for globular clusters is typically $f \sim 0.1$, while open clusters tend to have a higher binary

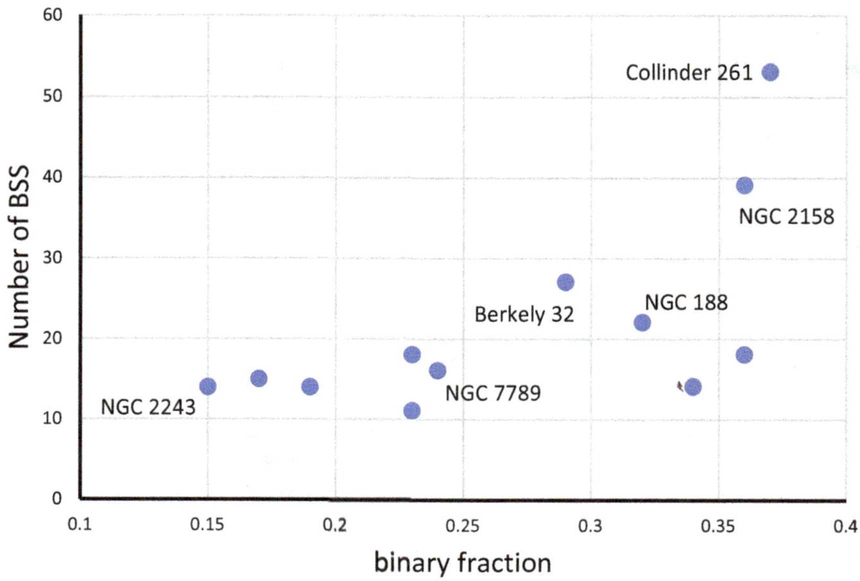

Figure 2.10. Number of blue stragglers within an open cluster plotted against the cluster's estimated binary fraction. A selection of cluster names has been provided for reference within the text. Data from [7].

fraction with (typically) $f > 0.3$. For all this, other environmental factors are important; for example, it appears that denser (that is, more compact) clusters have fewer binary systems compared to more loosely assembled clusters of the same mass and age. Figure 2.10 shows the result of comparing the number of blue straggler stars within a given open cluster, against the estimated binary fraction for that cluster. A general trend is again reasonably clear to the eye, in that the greater the binary fraction, the greater the number of blue stragglers observed. This result is highly suggestive, and supports the notion that at least one channel for blue straggler formation is that mediated through the evolutionary process operating in close binary systems. For all that, however, there is also evidence in Figure 2.10 for a complex cluster age, as well as cluster structure effect at play. The cluster Collinder 261, for example, is the oldest cluster in the sample, being some 7.9 billion years in age. The youngest cluster in the sample, however, is NGC 7789, being 1.6 billion years old.

Clusters NGC 2158 and NGC 2243 are about the same age (some 3.2 billion years), and yet they have distinctly different binary fractions and blue straggler populations. In terms of cluster characteristics, however, there is a distinct difference in their total masses, with NGC 2243 being some 4 times less massive than NGC 2243, and with NGC 2243 also being much more metal poor (recall chapter 1.2) than NGC 7789. How exactly the various cluster parameters (age, total cluster mass, core mass, binary fraction, metallicity, and compactness) play into the modulation of blue straggler numbers is, as yet, unclear (which is not to say that many researchers haven't attempted to quantify the relationships — see below).

2.9. Globular clusters

In addition to counting the number of blue stragglers in a given cluster, one can also look at how they are distributed within the cluster. That is, are the blue stragglers distributed at random throughout the entire cluster, or are there more blue stragglers in the central core, where the star density is highest, or, in contrast, are they more commonly located in the outer cluster regions, where the star density is lower? This is something that is particularly important with respect to globular clusters, where it is found that the majority of blue stragglers are located in the central regions (the cluster core). More than this, however, it is also apparent that in many clusters (but not all) the number of blue stragglers, while decreasing outwards from the cluster center, picks up again in the outermost regions. The population distribution of blue stragglers, radially speaking, is two two-peaked. Figure 2.11 shows the results of a 2018 study by Michela Mapellie (Centre for Astronomy at Heidelberg University) and co-workers [8], who looked at the relative number of blue stragglers (compared to the number of post-main sequence stars) as a function of radial distance from the center, in the globular clusters 47 Tucani, NGC 6752, and M 3. The quantity r_{av} is a normalizing factor between clusters, such that the minimum in the observed number of blue stragglers is always at $r / r_{av} = 1$. Here, a two-peak distribution is revealed. Not only is there an initial peak at the cluster center ($r / r_{av} \approx 0$), followed by a

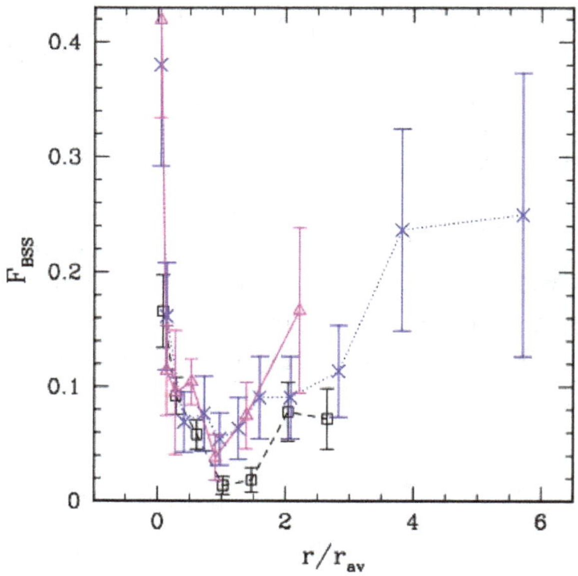

Figure 2.11. Radial distribution of blue stragglers (normalized to post-main sequence star counts) in three globular clusters: 47 Tuc (crosses), NGC 6752 (triangles), M 3 (squares). The radial distance is scaled so that the location of the minimum for the number of blue stragglers in each cluster falls at $r / r_{av} = 1.0$. Diagram from [8].

steep decline to $r / r_{av} = 1$, but a steady rise for $r / r_{av} > 1$ to a second, outermost maximum. What does this two-peak distribution tell us? Well, what it hints at is that there are at least two channels by which blue stragglers are produced within globular clusters. One mechanism dominates in the cluster core, while the other dominates in the cluster's outer most reaches. To probe deeper into this multi-channel blue straggler formation idea, we must first say a little more about the distribution and dynamics of stars within globular clusters.

Globular clusters are deceptively complex. They are not simply self-gravitating structures of stars that chanced to form, at the same time, and orbit about each other in a relatively small volume of space. This being said, the simultaneous formation part is taken as a fundamental property, and while the steps in the formation of globular clusters are still debated, the observations do indicate that the stars within such clusters really did form at the same time — some clusters, it should be said, do show evidence for multiple populations being present, but, for all that, the spread in population ages is small compared to the (present and ever increasing) age of the cluster. That globular clusters must be old, first-formed structures born when the galaxy itself was young and forming is clear from the spectra of their constituent stars, which are all low-mass and metal-poor. No globular cluster has ever displayed any observed evidence of ongoing star formation, and this indicates that the efficiency of star formation must have been high — this statement being made in contrast to star formation within the spiral arms of galaxies, where star formation is a relatively inefficient process. One of the key characteristics of globular clusters is their high packing of stars per unit volume of space. On average a star density of about 0.4 stars per cubic parsec is deduced, but this can increase from 100 to 1,000 stars per cubic parsec in a cluster's core. Recall from Chapter 1 that in the solar neighborhood the density of stars is about 0.1 stars per cubic parsec. Indeed, the number of stars per unit volume of space within a globular cluster decreases outward from the center, and the properties of a cluster are generally given in terms of three characteristic radii. These are the core radius, R_c, the half-light radius, R_h, and the tidal radius, R_t. These radii are derived according to measurements of surface brightness across the cluster. Figure 2.12 illustrates the surface brightness profile for the globular cluster Omega Centauri (recall figure 1.20). In this diagram, the y-axis indicates the energy output per second (the luminosity) per parsec squared as a function of the radial distance (plotted on the x-axis) from the cluster center. All globular clusters show a similar brightness profile, and this indicates that most of the energy output in the form of starlight is from the central regions, where, indeed, the number density of stars is the highest. Interior to the core radius, R_c, the surface luminosity is essentially constant — in the case of Omega Centauri, where $R_c \approx 4$ pc. The radius of half-light contribution, R_h, is located beyond the core radius, and for Omega Centauri this is at $R_h \approx 7$ pc. The outermost radius, R_t, is the tidal radius; this essentially indicates the outer boundary of the cluster and corresponds to the radial distance from

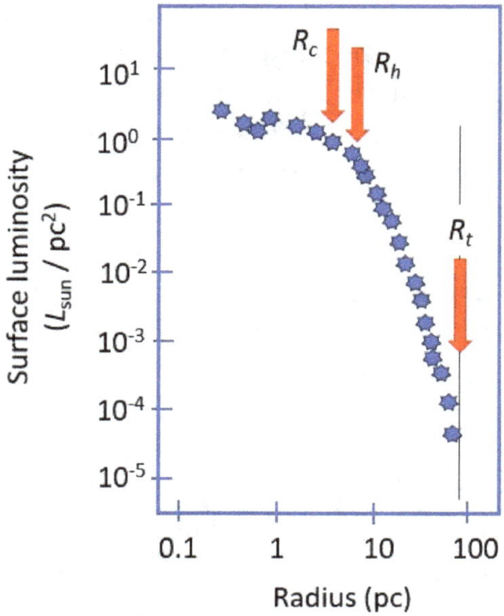

Figure 2.12. Surface brightness profile of globular cluster Omega Centauri. The vertical red arrows indicate the core, half-light, and tidal radii associated with the cluster. Image adapted from [9].

the center of the cluster at which the gravitational influence of the rest of the galaxy equals the gravitational influence of the cluster itself — for Omega Centauri this is located at a radial distance of $R_t \approx 68$ pc. With this information we deduce that the number of stars per unit volume of space must decrease outwards from the core since half of the total surface brightness is concentrated within an inner sphere of radius $R_h \approx 7$ pc, while the remaining half of the total surface brightness is produced in the annular shell stretching between $R_h \approx 7$ pc and $R_t \approx 68$ pc.

The central concentration of stars within globular clusters is a consequence of dynamic friction. That is, the close proximity of stars to each other within a globular cluster results in many close gravitational encounters, even collisions, and this tends to rob the more massive stars of their kinetic energy, causing them to 'sink' towards the cluster's central region. In contrast, the less massive stars tend to gain kinetic energy through gravitational interactions, and they will tend to migrate outwards from the center. This process of mass segregation is known as core collapse, and according to the initial density of stars in the cluster, it is a process that takes billions of years to come about. Further away from the core, where the star density is lower, encounters and interactions are less common, and accordingly these stars suffer less dramatic migration. Indeed, close to the outer regions of the cluster (i.e., with $R \approx R_t$) the gravitational influence of the galaxy drives an evaporation process, whereby stars are lost into the galactic halo.

The process of cluster core collapse is modified by the presence of binary systems, and this has importance with respect to the formation of blue stragglers. Initially, it is those stars located close to the cluster center that experience the greatest number of gravitational encounters, and begin, thereby, to experience core collapse. Indeed, in this early phase, binary systems are either ejected from the cluster or disrupted entirely, adding, thereby, to the 'crush' of stars in the core region. Following the initial core collapse, however, interactions between single stars and binary systems results in the single star gaining energy, with the two stars in the binary system becoming more tightly bound (that is, they end up closer together). The combined effect of these interactions results in the outer regions of the cluster expanding, and the appearance of a distinct, central population of compact binary systems. The blue straggler connection, as will be seen in the next chapter, is directly linked to the fate of those stars in the inner core, where direct collisions between stars are possible, and through the evolutionary pathways that are opened up to those stars locked within compact binary systems. Indeed, these latter compact binary systems produce a whole array of stellar exotica: not just potentially blue stragglers, but low-mass X-ray binary systems, millisecond pulsars, and cataclysmic variables. The last three of these exotic systems are related to compact binary systems containing either a white dwarf or a neutron star component.

That astrophysical exotica can lurk within the core regions of globular clusters has recently been demonstrated by Maximilian Häberle (Max Planck Institute for Astronomy, Heidelberg) and co-workers [10]. Indeed, these researchers have argued that an

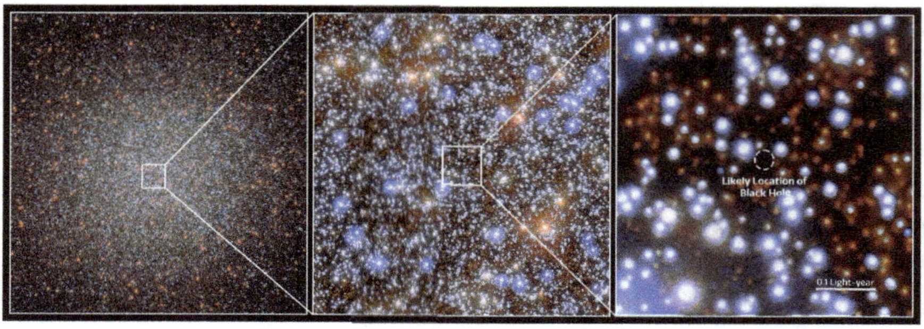

Figure 2.13. Zooming in (from left to right) towards the core of Omega Centauri (see also Figure 1.20). Using more than 500 images of the core region, extracted from the Hubble Space Telescope archive (accumulated over a 20-year time span), the proper motion tracks of seven central stars indicate that they have velocities greater than the cluster's escape velocity. These stars provide good evidence for the presence of an intermediate black hole at the cluster center (far right image). Image courtesy of ESA/Hubble, NASA, Maximilian Häberle (MPIA).

intermediate-mass black hole, with a mass of at least 8,200 $M_\odot$, exists at the center of Omega Centauri. Their analysis is a classical one, and was gained from a set of observations collected over a 20 year time interval. In particular, they followed a proper motion survey (recall Chapter 1.4) of stars in the central region of the cluster, and discovered seven stars that are moving with speeds in excess of the escape velocity of the cluster (Figure 2.13). Stars moving faster than the escape velocity of a cluster are most easily explained upon the basis that they are anchored in orbit about some massive object, and this is where the black hole hypothesis comes in. The exact orbits for the seven rapidly moving stars found by Häberle *et al.*, are not, as yet, known, since this is ongoing work. A second recent 2024 study of the central mass distribution of Omega Centauri by Andrés Bañares-Hernández (Instituto de Astrofisica de Canarias) and co-workers [11], using data relating to stellar motions and the accelerations deduced for millisecond pulsars, suggests that the situation at the core of Omega Centauri is even more complicated than that envisioned by Häberle *et al.* Indeed, Bañares-Hernández *et al.* argue that their data suggests an extended central mass distribution, possibly made up of multiple stellar-mass black holes. Where this discovery becomes relevant with respect to the possible formation of blue stragglers will be revealed in Section 3.18.

2.10. The daughters of Atlas

The closest open cluster to the Sun at the present epoch is the Hyades. Named after the five daughters of Atlas, the Hyades is set, in plain sight, amidst the constellation of Taurus (Figure 2.14). Located about 47 parsecs away, the cluster is estimated to be some 630 million years old. It has an estimated total mass of some 400 $M_{\odot}$, and a turn-off mass of $M_{TO} = 2.3\ M_{\odot}$. The tidal radius of the cluster, set by gravitational interactions with the galaxy, is set at about 10 pc, while the cluster has a distinct inner core radius of some 3 pc. Indeed, the cluster shows a distinct mass segregation effect, with the more massive stellar members being located within the core, and the less massive stars in the outer halo regions. Indeed, it is also thought that

Figure 2.14. The Hyades open cluster. The bright star to the left is Aldebaran (α Tauri — not actually a member of the Hyades cluster). The circled star 68 Tauri is the second closest blue straggler to the Sun, and the hottest star within the Hyades cluster. Image adapted from [12].

considerable star evaporation has taken place in the Hyades, with its initial total stellar mass being at least twice that seen at the present time. The Hyades stars are more metal-rich than the Sun and stars in the solar neighborhood, and interestingly (see chapter 4 later) 5 stars in the Hyades have been found to support exoplanetary systems. Additionally, a number of white dwarfs have been found in the cluster, and several of these have been found to show evidence for surface 'pollution' resulting from the accretion of rocky (that is asteroidal) material. This is suggestive that these evolved, now dead, stars once had terrestrial planetary companions [12].

Figure 2.15 shows the HR diagram for the Hyades as deduced from data collected at the Observatoire de Haute–Provence, France [13]. The isochrones shown in Figure 2.15 not only indicate an age in the order of 625 million years, but the presence of perhaps six blue stragglers beyond the turn-of mass. By far the most dramatic blue

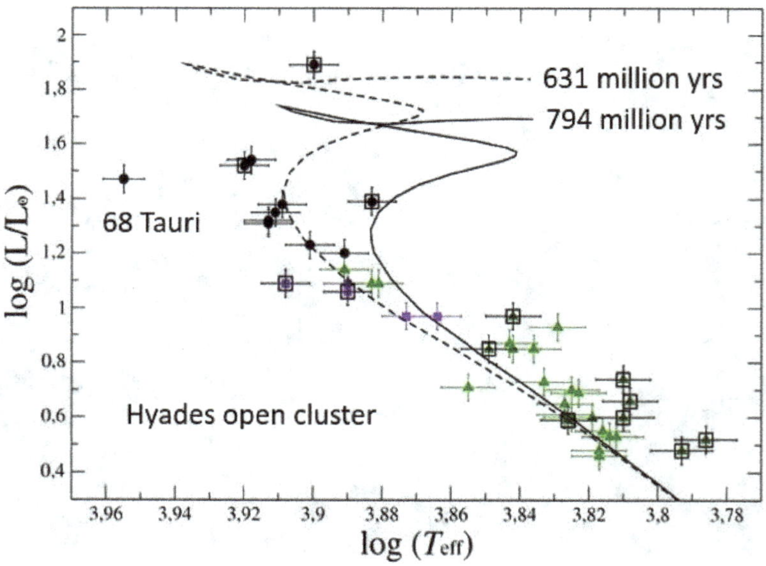

Figure 2.15. HR diagram for the Hyades open cluster. The black dots indicate A spectral type stars, the green tringles correspond to F spectral type stars, and the squares indicate binary systems. The blue straggler 68 Tauri is distinctive, being the hottest and 4th most luminous star in the Hyades cluster. Image adapted from [13].

straggler candidate is the star 68 Tauri (HD 27962). Indeed, this is the hottest star within the cluster, and with an apparent magnitude of 4.3, it is visible to the naked-eye[19]. Intriguingly, an Am spectral type has been deduced for 68 Tauri, indicating that it has a peculiar metal rich composition[20]. Additionally, Gerald Kuiper discovered in 1937 that 68 Tauri has a faint and close-in binary companion. That 68 Tauri, a distinctive blue straggler, has a peculiar composition as well as a close companion is highly suggestive of it having a complex evolutionary history (see Chapter 3). The closest (field) blue straggler to the Sun, at a distance of just 14 pc, is contained within the triple-star system α Fornacis. In this system, the secondary star α For Ba has a significantly enhanced barium abundance over that of the primary α For A. The reason for this enhancement is thought to be a result of mass transfer with a third companion star α For Bb. This latter star, the mass donor, is now a close-in, white dwarf companion to α For Ba.

2.11. A summary of sorts

> *Science seeks knowledge. Let the knowledge lead us where it will, we still must seek it.*
>
> (Professor Challenger in *The Day the Earth Screamed*)

Walt Whitman, in his famous poem *When I Heard the Learn'd Astronomer* (recall footnote 2 in Chapter 1), commented that he soon tired of charts and tables, but then, he was no astrophysicist. Indeed, to the astrophysicist, charts, diagrams, and ordered columns of numbers are the clay from which they make their bricks. So, even though we haven't, as yet, addressed all the observational characteristics and possible origin avenues for blue stragglers, we shall finish this chapter with a few more general statements from the observations and present a table with the columns arrayed before you.

[19] This star has many designations, being additionally cataloged as HR 1389, ADS 3026, and delta-3 Tauri. It is also traditionally known as Cleeia, one of the five Hyades nymphs, and daughter of Atlas.

[20] The bright star Sirius (indeed, the brightest in our sky — apart from the Sun) is also classified as an Am star.

Table 2.2. Summary table of gross blue straggler characteristics according to where they are found — globular cluster (second column), open cluster (third column) and outside of a cluster (fourth column).

	Globular clusters	**Open clusters**	**Field**
Distribution	Cluster center Sometimes bi-modal	Cluster center Sometimes bi-modal	Isolated objects
Percentage	0.01 to 0.001	1 to 0.1	Unclear
Binarity	Unclear (~ 50% ?)	High (~ 75%)	High (100% ?)

If one thing is clear about blue stragglers, it is that they are commonly (but not exclusively) found with binary systems. The companion stars can be a main sequence star, a helium burning star, or even a white dwarf. Some blue stragglers have accompanying discs about their equators, others do not. In terms of their rotation, some blue stragglers spin much faster than their main sequence counterparts; others spin more slowly. Typically, the surface composition of blue stragglers is the same as that seen in their cluster companions, but in some circumstances, significant compositional differences are observed. And finally, Table 2.2 summarizes the typical characteristics of blue straggler stars with respect to where they reside, their percentage of the total population, and their typical binary fraction.

Come, Watson, come! he cried. The game is afoot.
Not a word! Into your clothes and come!

(*The Abbey Grange*)

References

[1] M. Beech. Since when was the Sun an ordinary star? *Journal of the Royal Astronomical Society of Canada*, 105, 232–227, 2011.

[2] W. H. McCrea. *Physics of the Sun and Stars* Hutchinson's University Library, London, (1950).

[3] F. Adams., and G. Laughlin. *The Five Ages of the Universe*. Free Press Publishers, New York. 1999.

[4] M. Beech. *Introducing the Stars: formation, structure and evolution*. Springer Nature, Switzerland AG. 2019.

[5] E. Burbidge., and A. Sandage. The color-magnitude diagram for the galactic cluster NGC 7789. *The Astrophysical Journal*, **128**, 174–184, 1958.

[6] W. H. McCrea. Extended main-sequence of some stellar clusters. *Monthly Notices of the Royal Astronomical Society*, **128**, 147–155, 1964.

[7] M. Rain., A. Ahumada., and G. Carraro. A new, Gaia-based, catalogue of blue straggler stars in open clusters. *Astronomy and Astrophysics*, A67, 2021. https://doi.org/10.1051/0004-6361/202040072.

[8] Michela Mapellie., *et al.* The radial distribution of blue straggler stars and the nature of their progenitors. *Monthly Notices of the Royal Astronomical Society*, **373**, 361–368, 2006. https://doi.org/10.1111/j.1365-2966.2006.11038.x

[9] P. Raffaele., J. Binney., C. Nipoti., and L. Posti. Action-based models for dwarf spheroidal galaxies and globular clusters. *Monthly Notices of the Royal Astronomical Society*, **488**, 2423–2439, 2019. https://doi.org/10.1093/mnras/stz1617.

[10] M. Häberle., *et al.* Fast-moving stars around an intermediate-mass black hole in ω Centauri. *Nature*. 10 July, 2024. DOI: 10.1038/s41586-024-07511-z

[11] A. Banares-Hernandez., F. Calore., J. Camalich., and J. Read. New constraints on the central mass contents of Omega Centauri from combined stellar kinematics and pulsar timing. *Astronomy and Astrophysics*, 693, A104, 2025. https://doi.org/10.1051/0004-6361/202451763.

[12] From Overview of the Hyades Star Cluster: https://sci.esa.int/web/hubble/-/51756-overview-of-the-hyades-star-cluster-ground-based-image.

[13] M. Gebran., M. Vick., R. Monier., and L. Fossati. Chemical composition of A and F dwarfs members of the Hyades open cluster. *Astronomy and Astrophysics*, 523, A71, 2010.

Chapter 3
Making blue stragglers

Malingering is a subject upon which I have sometimes thought of writing a monograph

(*The Dying Detective*)

Now that the existence of blue stragglers as a real astrophysical phenomenon has been established, our attention accordingly shifts to the question of how it is that such stars have formed. Is there just one way to make a blue straggler, or are there many ways? Indeed, what determines the number of blue stragglers that might exist within a given star cluster, and what do their observable characteristics tell us about their internal structure?

Figure 3.1 shows the various pathways or channels that have been presented for the formation of blue stragglers. We shall look at each of these channels, in turn, as we move through this chapter, except, that is, for the channel identified as 'Other' — this channel will be the main focus of Chapter 4. The currently favored blue straggler production channels are those involving collisions and mass transfer. The delayed and mixed formation channels are taken to be possible, but generally unlikely, or, at best, very low efficiency pathways.

The key to solving the blue straggler origins mystery, along the lines illustrated in Figure 3.1, is the identification of mechanisms that can make a star appear to be evolving more slowly through its main sequence phase than an identical-mass, concomitantly born

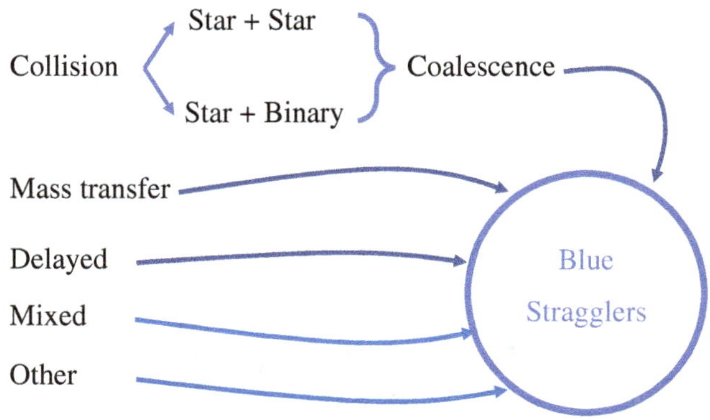

Figure 3.1. Evolutionary channels for the production of blue stragglers. The 'Other' channel will be the subject of Chapter 4.

companion. In terms of timescales, a blue straggler in a specific star cluster satisfies the inequality that $T_{\text{ms}}(M_{BS}) > T_{\text{ms}}(M_{\text{TO}})$, where $T_{\text{ms}}(M)$ is the main sequence lifetime of a star of mass M. This is in contrast to standard evolution theory, which posits that $T_{\text{ms}}(M_{BS})$ should be less than $T_{\text{ms}}(M_{\text{TO}})$, since the mass of the blue straggler is greater than its parent cluster's turn-off mass: $M_{\text{BS}} > M_{\text{TO}}$. Recall, this is required on the basis that the observed luminosity and surface temperatures of blue stragglers are higher and hotter than those corresponding to a star of mass M_{TO}. The trick then, so it seems, is to find ways of adding mass to a star initially close to the turn-off mass, and in the process, reset its evolutionary clock.

3.1. Delayed star formation

We would not dare to conceive the things which are really mere commonplaces of existence.

(*A Case of Identity*)

Before considering ways of adding mass to a star perhaps we should tidy away the most obvious explanation for the origin of blue stragglers. Indeed, here the argument is simply that the blue stragglers are entirely ordinary stars which formed after the bulk of other stars in the cluster. In this sense, a secondary or delayed birth process is called into play. It is an entirely reasonable idea, and it carries the benefit of not requiring the invocation of a hybrid form of single star evolution. If the turn-off mass for the cluster is taken to be M_{TO}, then the cluster age will be of order $\tau_{cl} = 10^{10} /(M_{TO})^{2.5}$ (years), where the mass M_{TO} is expressed in solar mass units. Likewise, the main sequence lifetime of a star of mass M is $\tau_{ms} = 10^{10}/M^{2.5}$ (years). Accordingly, if we define $\Delta t = \tau_{cl} - \tau_{ms}$, then

$$\Delta t = \tau_{cl}\left[1-\left(\frac{M_{TO}}{M}\right)^{2.5}\right] \tag{3.1}$$

In this manner, if $M = M_{TO}$ then $\Delta t = 0$, which makes sense since the star with mass M will have the same main sequence lifetime as the turn-off mass for the cluster — i.e., it must have formed at the same time as the stars having the turn-off mass. If, however, for example, $M = 2\ M_{TO}$ then $\Delta t = 0.82\tau_{cl}$, or, in other words, the main sequence lifetime of the star must be less than $0.18\tau_{cl}$. Accordingly, if the cluster is, say, $\tau_{cl} = 1$ billion years old (indicating $M_{TO} \sim 2$ $M_\odot$ — recall Equation 2.7), then $t_{ms} \sim 180$ million years, indicating $M \sim 4\ M_\odot$. Alternatively, if, say, $M = 6\ M_{TO}$, then $\Delta\tau = 0.99\ \tau_{cl}$, or $t_{ms} \sim 0.01\tau_{cl}$, and for our 1-billion-year-old cluster this means that $M \sim 12.5\ M_\odot$ (i.e., a star with main sequence age $t_{ms} \sim 10^7$ years. What all this implies is that the more massive a blue straggler is relative to the turn-off mass in a given cluster, the more recently in the history of that cluster it must have formed. The other way of saying this is that the cluster must have been producing stars over most of its lifetime. To produce a blue straggler of mass $M = 2\ M_{TO}$ the cluster must have been producing stars throughout at least 82 percent of its history. The problem with this conclusion, however, is that there is absolutely no evidence that clusters, globular or galactic, are capable of producing stars over such extended periods of time. To do this a cluster would have to show some evidence of gas reserves as well as regions undergoing star formation. Problematically, however, not one known globular cluster has ever been observed to show any evidence of ongoing star formation. Likewise, there is no evidence for ongoing waves of star formation within open clusters. The reason for this latter conclusion is based upon the idea that the gas and dust needed to produce later populations of stars are quickly driven out of any newly emerging galactic cluster by the formation of massive stars. These massive stars support strong stellar winds, have very short main sequence lifetimes of just a few millions of years, and rapidly undergo supernova disruption. The effect of all this is to rapidly drive any residual gas and dust not incorporated into the first generation of stars out of the cluster environment, thereby quenching any subsequent star formation.

While seemingly the most obvious solution to the blue straggler mystery, the delayed formation scenario is, in fact, the least likely way in which such stars might have formed. Yes, blue stragglers are delayed in their evolutionary appearance, but it is not because they are significantly younger (that is, formed later) than the other stars within their host cluster.

3.2. Accretion from the ISM

Forward, forward let us range.
Let the great world spin for ever down the ringing grooves of change
(A. L. Tennyson: *Locksley Hall*)

One way in which a star might acquire additional mass after its formation is by the accretion of gas from the interstellar medium. In this manner, as a star moves through a gas cloud, some of the gas will be accreted at its surface, thereby increasing its mass and making it more luminous — in this sense, making it move up the main sequence in the HR diagram. Could it be that blue stragglers are those stars within a cluster that, purely by chance, have captured an additional amount of interstellar material as the cluster aged? The accretion question was, in fact, considered in great detail by Hermann Bondi, Fred Hoyle, and Raymond Lyttleton during the 1940s and 50s, and they reasoned that for a star of mass M moving with speed v through an interstellar cloud of density ρ, that the accretion rate will be of order

$$M_{acc} = \frac{4\pi G^2 \rho M^2}{v^3} \tag{3.2}$$

where G is the gravitational constant. Maximum accretion rates occur, therefore, when a star moves slowly through a high-density cloud. The characteristic density of a typical star-forming interstellar cloud is $\rho \sim 10^{-19}$ kg/m^3, and for a star moving through such a region at a relative velocity of 5 km/s, the accretion rate will be of order $M_{acc} \sim 10^{-12}\ (M\ /\ M_\odot)^2$ solar masses per year. With this estimate in hand, if a solar mass star was to accrete matter for its entire main sequence lifetime, then its mass might increase by about 0.1 percent. This is not enough matter to significantly change its evolutionary history. Additionally, continuous accretion over a timescale of billions of years is simply not tenable. Star forming regions (which have the highest-density characteristics of the interstellar medium) are typically only a few parsecs across — at 5 km/s, the crossing time of such a cloud is in the order of 400,000 years, which is very, very much shorter than the main sequence lifetime of a Sun-like star, or indeed, any star initially less massive than about 60 $M_\odot$.

The idea of forming a blue straggler, through the accretion of interstellar material, is simply not viable. Essentially, the interstellar medium is not dense enough for the accretion of material to be either efficient or significant — even on timescales as long as the main sequence lifetime of star. If a star is to gain an appreciable amount of mass after it has formed, then some different, and faster-acting mechanism is needed.

3.3. All mixed-up

A good detective knows that every task, every interaction, no matter how seemingly banal, has the potential to contain multitudes.

(*Elementary* — CBS TV series)

There were no electronic computers capable of solving the equations of stellar structure in 1953 when Sandage first found evidence for blue straggler stars. There were researchers, however, making calculations on mechanical, hand-driven calculators, and these researchers, for obvious reasons, made their calculations as simple as possible. Indeed, it could take several days of adjusting mechanical keys and turning a hand-crank to produce a single model sequence, and accordingly, few researchers had the time (and/or the patience) to construct extensive sets of evolutionary calculations. Furthermore, at this mid-19th century time point, it was entirely unclear if stars were fully mixed within their interiors, — having, thereby, a near homogeneous composition throughout — or whether there was a distinct compositional difference between the core and envelope (that is, that their interiors were inhomogeneous). It was somewhat easier, in a time-saving sense, to construct models that were homogeneous, and these calculations indicated that as a fully mixed star evolved, it became hotter and more luminous, essentially moving upwards and along the main sequence — exactly the effect required to explain the existence of blue stragglers.

The effect of mixing within a star is illustrated in Figure 3.2. In this case, a series of detailed numerical models have been constructed for a 2 $M_\odot$ star with varying amounts of internal mixing being artificially included. The amount of mixing is set according to a parameter called q_{mix}, which is the mass fraction of the model that is taken to be chemically homogeneous. The models clearly indicate that as q_{mix} increases towards its upper limit of 1, where the whole star would be fully mixed with a homogeneous composition, the evolution moves upwards along the main sequence — that is, towards a higher luminosity and temperature — exactly as needed to explain blue stragglers. Not only do the mixed models fall in the right part of the HR diagram to explain the observed luminosity and temperature characteristics of blue stragglers, they also explain their long lived, or time-straggling nature. This is illustrated by the tick marks on each track in Figure 3.2. The main sequence lifetime of a non-mixed 2 $M_\odot$ star is (as indicated by the lower track in Figure 3.2) about 7×10^8 years, while that for a 2 $M_\odot$ star with $q_{mix} = 0.95$ (upper and leftmost track in figure 3.2) is about 20×10^8 years — nearly three times greater. Mixing a star makes it evolve more slowly — indeed, extensive internal mixing is the astrophysicists equivalent of the philosopher's stone. All this, of course, makes physical sense: by forcing the mixing of material across a large mass fraction of a star's interior, the star has access to a much greater

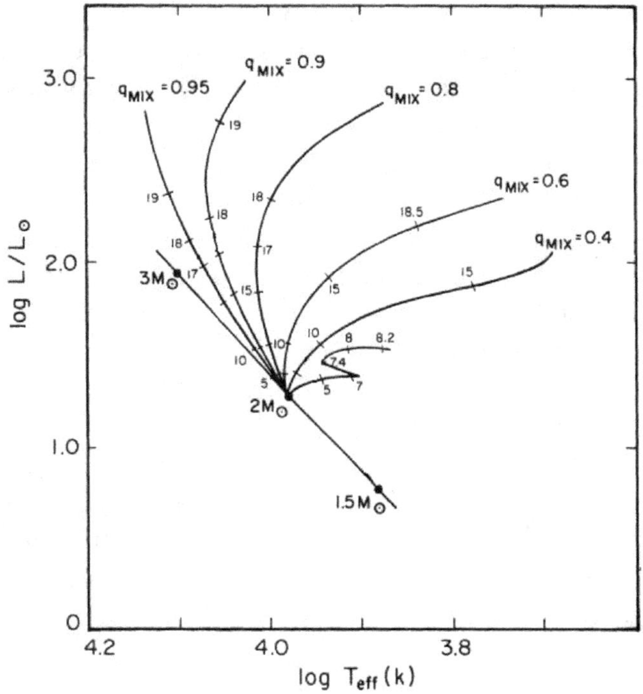

Figure 3.2. The effect of forced internal mixing upon a star's evolution. The q_{mix} term indicates the interior mass fraction that is mixed, and the tick marks on each evolutionary tract indicate the passage of time in units of one-hundred million years. The calculations end at the moment of core hydrogen exhaustion. Image adapted from [2].

reservoir of hydrogen than it would otherwise have had under canonical conditions. Could it be, as J. Craig Wheeler (University of Texas, Austin) questioned in 1979, that the origin of blue stragglers "is the [result of] funny business in the main sequence?" [1]

Access to a greater reserve of hydrogen will prolong the main sequence lifetime of a star. Furthermore, we can estimate the extent of mixing that is minimally required by revisiting the formula, derived in Chapter 2, for the nuclear timescale τ_{nuc}. Looking to Equation (2.4), the term of interest now is the corrective f term. This term was set to $f = 0.1$ on the basis of comparisons with detailed numerical model calculations, which indicate that the main sequence lifetime is completed once the inner 10% of hydrogen within a star's central core is exhausted (that is, converted to helium through fusion reactions). In this manner, letting the term be a variable quantity, the nuclear timescale — which, recall, is the canonical main sequence lifetime — can be rewritten as $\tau_{\rm nuc} = t_{\rm ms} = f \times 10^{11}/M^{2.5}$ years, where M is expressed in units of the Sun's mass. Now, consider the formation of a star cluster. We take it that all the stars formed at the same time, but we are now invoking the possibility that some stars (for whatever reason) are more 'mixed' than others — such stars will be observed as blue stragglers. By "more 'mixed'" it is simply intended that they have a larger reservoir of hydrogen that can be mixed into their cores — that is, larger f-values than in the canonical case, where $f_{normal} = 0.1$. Let us look at the situation where we have a star that is two times more massive than the cluster turn-off mass, but we require that it have the same main sequence lifetime appropriate to that of the cluster turn-off mass (that is, $t_{MS,mixed} = t_{MS,TO}$). Accordingly:

$$\frac{t_{MS,mixed}}{t_{MS,TO}} = \frac{f_{mixed}}{f_{noraml}}\left(\frac{M_{TO}}{M_{mixed}}\right)^{2.5} = 1$$

Setting $M_{mixed} = 2\ M_{\rm TO}$, and $f_{normal} = 0.1$, we find that $f_{mixed} = 0.56$. That is, a star having twice the mass of the cluster turn-off mass will have the same main sequence lifetime as a turn-off mass star if it can tap some 56% of its internal hydrogen. In this manner, if a single star is to become a blue straggler through additional mixing, then it must be pretty well mixed. Indeed, well mixed throughout its main sequence evolutionary phase.

With fully- or largely-mixed evolutionary models we begin to make some progress towards explaining the existence of blue stragglers, but we have replaced one problem with another: why are only a very few stars within a given cluster evolving in a mixed-up manner. In the case of Sandage's observations of M 3, only about three percent of the stars that he studied appeared as blue stragglers. Indeed, blue stragglers are both rare as well as exotic. At best, this result informs us that most stars are not well-mixed — at least on the main sequence while converting hydrogen into helium[1]. What then, might stir up the interior of a star so that it develops a near homogeneous chemical composition? Certainly, energy transport by convection will result in chemical homogeneity, but the domain over which convection proceeds need not mix material into the central core — in the Sun, for example, only the outer third is in convective motion. Accordingly, deep mixing during the main sequence phase requires the invocation of some physical mechanism that is typically not important within the main sequence evolutionary story of most stars.

The evolutionary tracks shown in Figure 3.2 were calculated by Hideyuki Saio and J. Craig Wheeler in 1980 [2], and they are an example of numerical experimentation. The models are based upon the full equations of stellar structure, with the one exception of invoking chemical homogeneity at some set q_{mix} value within the interior. In other words, the chemically homogenizing mechanism is imposed by hand, as it were, and is not based upon some specific, naturally operating physical process. This, of course, is a perfectly fine and acceptable way of proceeding, and simply adopts the philosophy that if the homogenized models provide a better explanation for the appearance of blue stragglers than the canonical models, then this suggests an explanation for what is actually going on, and the next task is to identify the physical mechanism at play. Indeed, astronomers wrestled with this sort of problem for many years prior to the development of electronic computers. As indicated in Chapter 2, what was at stake was literally time. Time, that is, during which the astronomers would have to crank out calculations by hand. Accordingly, whenever solutions to the equations of stellar structure and/or evolution were needed, the question was how to simplify the calculations so that answers could be obtained in a reasonable amount of time — time in this case being measured in days and weeks. In this manner, the choices available to researchers were to either simplify the physical input and apply mathematical rigor, or to simplify the mathematics and maintain physical rigor. These

[1] Recall from Chapter 2, however, that low mass stars are naturally well-mixed since energy is transported by convective motion within the greater part of their interiors.

two approaches could, hardly surprisingly, produce distinctly different results, and the question remained as to which set of results one should believe. The first calculations of stellar structure simply assumed chemical homogeneity, and it was not until the late-1930s that Ernst Öpik (Tartu Observatory, Tõravere) first considered the effect of allowing for the build-up of a core-envelope composition discontinuity. Fred Hoyle and Raymond Lyttleton at Cambridge University, along with Martin Schwarzschild at Princeton University, further developed such model calculations in the early- to mid-1940s and found that the chemically stratified model stars evolved in a totally different manner to the fully-mixed ones. The luminosity still increased as a non-homogeneous star model aged, but now it becomes much larger in physical size and much cooler at its surface — indeed, such model stars evolve into stars that fall in the red giant region of the HR diagram. From the 1960s, with the help of electronic computers[2], it became increasingly clear that the interiors of most stars are not, in general, extensively mixed during their main sequence phase[3]. So, once again, we come back to the problem that, if blue stragglers are the result of extensive internal mixing, then what processes are doing the mixing?

3.3.1. *That extra push*

> *Well, we can adopt it as a working hypothesis, and then see how far our difficulties disappear*
>
> (*The Valley of Fear*)

In addition to conducting numerical experiments with respect to mixing the interiors of stars, Saio and Wheeler [2] also investigated

[2] The emergence and general availability of electronic computers, in the late-1960s, freed astronomers from the drudgery of hand calculations, but it was still another 15 to 20 years before even electronic computers had the memory and speed capabilities to construct both physically and mathematically rigorous stellar models.

[3] There is good evidence, however, that stars more massive than about 100 $M_\odot$ do evolve in a largely homogeneous manner, and avoid the development of a red giant phase. This is largely a result of significant mass loss taking place, the presence of rapid rotation, and because of their high luminosity.

the effect of introducing a parameterized non-thermal pressure support term, δ. In this manner, the numerical experiments investigated one of fundamental properties of stellar structure. The very fact that stars can exist in a long-lived stable state (that is, have a main sequence phase at all) is the result of a dynamical balance being established. Gravity acts inwards in the sense that it continually tries to collapse a star down to a singularity, thereby producing a permanent black hole state. In contrast, an outward-acting internal pressure continuously tries to make a star as large as possible — dispersing the stellar gas to 'infinity' if left unchecked. We have, accordingly, two extreme stable states to which a star might trend: one in which a star has zero size (a singularity), and one in which it has an extremely large size (a dispersed cloud of interstellar gas). As with all things, however, a compromise is possible, and at each instant a star hovers in a dynamically balanced state, neither appreciably shrinking, nor appreciably expanding. To achieve this dynamic balance, the temperature of a star must increase inwards [3]. The idea here is that the outward acting pressure gradient is determined by the internal temperature gradient; if a star, at any point within its interior, is to support the weight of overlying layers, then the pressure must be at its highest at the star's center, and essentially zero at its surface. In terms of outward acting pressure terms, there is the thermal pressure, $P_{TH,}$ due to the fact that the star's interior is hot; the pressure due to radiation, P_{RAD}; and if the conditions are right (as discussed earlier with respect to white dwarfs and neutron stars), the degeneracy pressure term, P_{DEG}, due to the Pauli exclusion principle. All these pressure terms act to stop gravity (the weight of overlying layers) from crushing a star out of existence. In general, the outward acting pressure term, P, at any point within a star can be written as a sum of terms, with $P = P_{TH} + P_{RAD} + P_{DEG} + \delta$, where δ is the parameterized non-thermal pressure support term investigated by Saio and Wheeler. The δ term is a catchall parameter; it is generally taken to be zero, but with the understanding that there are circumstances where it might not be zero. Such circumstances coming about naturally through the presence of a strong internal magnetic field and/or rapid rotation. While such physical processes can be studied in great mathematical detail, the actions add significant difficulties to the general computational process, and it is much simpler (numerically speaking) just to parameterize the effect — hence the δ term.

The numerical experiments conducted by Saio and Wheeler [2] found that a star, in which a non-zero, non-thermal pressure support term ($\delta > 0$) had been imposed, evolved with much lower surface temperatures and at lower luminosities than those of a similar mass but with the $\delta = 0$ condition enforced. In addition, the model stars with $\delta > 0$ evolved with much lower central temperatures, and accordingly consumed their central hydrogen supply much more slowly than similar mass stars in which $\delta = 0$ was enforced. In this manner, stars with a non-zero non-thermal pressure support term evolved much more slowly, with the main sequence lifetime being greatly extended by a factor of 10 and more. Here, then, is a possible reason for the existence of blue stragglers: they are not late-born stars within a cluster, nor need they be fully mixed within their interiors. No, they are stars that have an additional, non-thermal, pressure support term at play within their interiors.

While, once again, a physically allowable mathematical loophole has been exploited to place a model star in a long-lived state that mimics that of a blue straggler, we still run up against a problem: why should some stars have a non-thermal pressure support term $\delta > 0$, when other stars (indeed the majority) do not? To this quandary there is no clear answer. It is entirely possible to invoke physical reasons (rotation and/or the buoyancy support of a magnetic field, for example) for the δ term being non-zero, but there is no clear answer as to why such reasoned mechanism should turn on and/or off in order to explain the existence of blue stragglers.

3.3.2. *Rotation and magnetic fields*

> *There is nothing more stimulating than a case where everything goes against you.*
>
> (*The Hound of the Baskervilles*)

That the Sun spins was first discovered by Galileo in the early 17th century. Observing the steady shift in the positions of sunspots across the disc of the Sun when projected onto a screen by a telescope, Galileo correctly reasoned that the dark spots were not objects (planets or moons) orbiting the Sun, but rather blemishes on the Sun's 'surface' that moved across the field of view because the Sun itself was spinning. This was a brilliant piece of reasoning, and indeed, the first astrophysical result about a star based upon direct telescopic observations[4]. In the modern era, the Sun's rotation is measured according to Doppler shift measurements of photospheric absorption lines. Such studies indicate a characteristic rotation period of 25 days. The Sun actually undergoes differential rotation, however, with the equatorial regions rotating more rapidly than the polar regions. This differential rotation is vital to winding up the Sun's surface magnetic field lines and driving its 11-year sunspot cycle[5]. The Sun's magnetic field is actually generated at the base of its convective envelope, located some 1/3 of the radius inwards from the photosphere, and the interior of this boundary (the so-called tachocline) rotates with a near constant velocity (like a solid ball).

That rotation can critically alter the structure of a star follows from a consideration of how the centrifugal force modifies the gravitational acceleration at each location within a star's interior. The effective gravitational acceleration term g_e, at each layer within a star is given by g_e = (gravitational acceleration – centrifugal acceleration). In mathematical form this can be written as

$$g_e = \frac{GM}{R^2} - \frac{V^2}{R} \tag{3.3}$$

where R is the distance from the center, M is the mass interior to radius R, and V is the rotational velocity [3]. At the surface of a star, Equation (3.3) can be used to define a critical spin velocity V_c, above

[4] This being said, in Galileo's time it was certainly not clear that the Sun was a star and vice versa.

[5] The sun spot cycle is actually highly complex, and is additionally modulated according to the 22-year magnetic reversal cycle.

which the star will literally begin to tear itself apart. In this critical case, $g_e = 0$, and we have a rotation velocity limit of

$$V_c = \left(\frac{GM}{R} \right)^{1/2} \tag{3.4}$$

Accordingly, for a given mass star, the critical spin velocity will increase the smaller the radius; coversely, the larger the radius, for a given mass, the slower the critical rotation velocity.

For the Sun the critical rotation speed is about 435 km/s (implying a rotation period of some 3 hours), which is well above its actual rotation velocity of some 2 km/s (rotation period of 25 days). For a white dwarf having the same mass as the Sun, but with a radius comparable to that of the Earth, the critical spin velocity is some 4,000 km/s, translating to a rotation period of about 13 seconds. For neutron stars, the critical spin rate is even higher, at some 100,000 km/s and a rotation period of one-thousandth of a second. Remarkably, white dwarfs and neutrons stars spinning with periods of tens of seconds and milliseconds respectively are known to exist.

Subcritical rotation condition appears to be the rule of thumb for main sequence stars (see Figure 3.3). The most rapidly rotating main sequence stars are those with masses in the range of 5 $M_\odot$ to 40 $M_\odot$, with spectral types between A and O. These latter stars characteristically have rotation velocities in the order of 200 km/s. As we saw in Chapter 2, on the main sequence, the radius scales approximately with the mass of the star: $R \sim M^{0.8}$. Accordingly, the critical spin velocity does not vary significantly with respect to mass (i.e., spectral type). For all this, lower mass stars, as indicated in Figure 3.3, do have significantly slower rotation velocities than their higher mass companions. There are several physical reasons for this. One important factor is indicated in Figure 3.3, and is concerned with the presence, or lack, of a deep surface convective zone. Stars less massive than about 2 times the mass of the Sun do have convective envelopes, and, just like the Sun, such stars support surface magnetic fields. Observationally, the presence of a magnetic field is betrayed through its effects upon the absorption lines within a star's spectrum (through the Zeeman effect) and by the generation of flares and star spots.

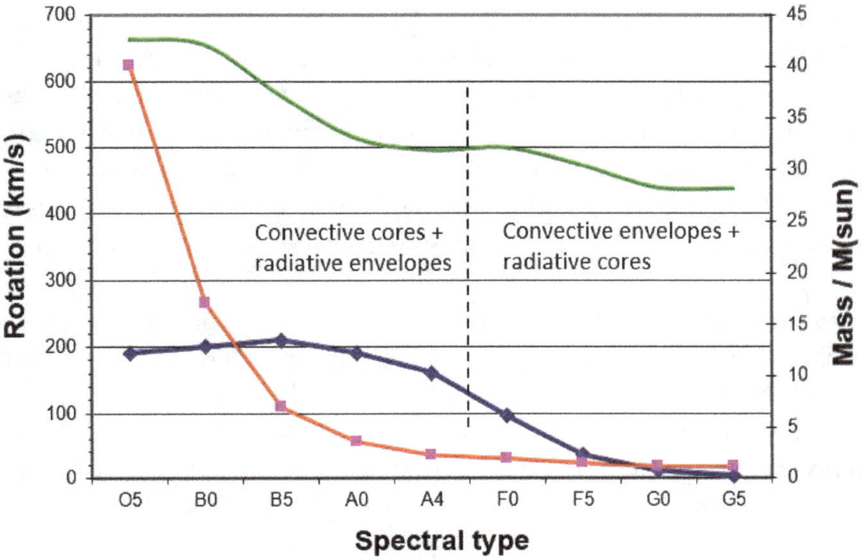

Figure 3.3. Rotation characteristics of main sequence stars. The blue curve indicates average observed rotation velocities, while the red curve indicates the mass corresponding to the various spectral types. The green curve is based upon Equation (3.4), and indicates the critical rotation velocity limit for each spectral type (or stellar mass). The vertical dashed line between spectral types A4 and F0 indicates the mass range in which stars change over from having radiative envelopes (stars more massive than about 2 $M_{\odot}$) to convective envelopes (stars less massive than about 2 $M_{\odot}$) — see text for details.

Stars more massive than about 2 times that of the Sun have radiative outer envelopes, and while a magnetic field might be generated in their central convective regions this does not automatically translate to the presence of a stronger surface magnetic field. Indeed, in order to generate a self-sustaining magnetic field, a star must have a region in which rotation and convective turn-over exist. Furthermore, there must also be a region in which the magnetic field can be anchored (the tachocline), else it will rapidly decay. By far the most magnetic of stars are neutron stars and white dwarfs, where the field in the former case can be so strong as to generate the pulsar and magnetar phenomena. All low mass (red dwarf) stars have surface magnetic fields, as evidenced by flare activity and the formation of star spots (the star equivalent of sunspots). Among main sequence stars, the A and B spectral types exhibit the strongest magnetic fields, with only about 10 percent of O spectral type stars showing any signs of magnetic activity. That O spectral type stars show any signs of magnetic activity is actually surprising since they have deep radiative envelopes where it is not possible for a self-sustaining magnetic dynamo to operate. For these stars, we are either seeing evidence of a primordial field — that is, a remnant of the star formation environment — or we are seeing the effects of a deep structural perturbation — caused, perhaps, by the impact of a lower mass star, or a merger event within a close binary system (as to be discussed further below).

The existence of a strong surface magnetic field becomes especially important, with respect to rotation, when a star is also undergoing mass loss. The Sun, for example, is losing mass via the solar wind at a rate of about 10^{-14} $M_\odot$/yr. A massive O spectral type star loses mass at an even higher rate, perhaps as high as 10^{-5} $M_\odot$/yr. Such stellar winds are the important factor in controlling the spin-down of lower mass stars. To see how this works, we first need to introduce the idea of angular momentum. For a given mass M rotating at a distance R from some specified spin axis, with a speed V, the angular momentum is defined as: $L = MVR$. The most fundamental aspect of angular momentum, however, is that it is a conserved quantity. That is, if some change happens to the system, say the radius R becomes smaller, the conservation of angular momentum requires that the rotation velocity must increase in order that L maintains a constant value. If the radius changes from R_1 to R_2, the velocity must change from V_1 to V_2 so that $L_1 = L_2$, and accordingly, $M\ R_1\ V_1 = M\ R_2\ V_2$. Additionally, if $R_1 > R_2$, then $V_2 > V_1$, i.e., the spin velocity must increase. In the case of a star undergoing mass loss, the presence of a strong surface magnetic field results in a drag force or braking term being applied to the star, slowing, in consequence, its rotation speed. Effectively, angular momentum is transferred from the star to the outflowing wind, and as the wind moves away from the star (increasing its R value), its rotation velocity must slow down to conserve angular momentum. To the star, it is as if it has to pull against the magnetic field anchored into the slowly rotating wind, and accordingly, it begins to gradually slow its rate of rotation. What happens next is a combination of competing factors. To explain the observed rotation curve in Figure 3.3, we need to look at initial mass, the mass loss rate, and the main sequence lifetime. For massive stars, we have high mass loss rates, but weak surface magnetic fields and short main sequence lifetimes. For low mass stars we have low mass loss rates, strong surface magnetic fields, and long main sequence lifetimes. The result of all this is that, in general, a massive main sequence star does not stay on the main sequence long enough for its rotation speed to be appreciably slowed. A low mass star, in contrast, remains on the main sequence for an extended period of time and can consequently undergo significant rotational slowing. Indeed, in low mass main sequence stars, the Skumanich Law (named after Andrew Skumanich, who first described the law in 1972) indicates that the rotational velocity for a given spectral type (e.g., Solar or G2 spectral type stars) decreases as the inverse square root of age, with older stars rotating more slowly than younger ones[6]. While we now have some handle upon the observations and expectations of how stars spin, that is, the various features indicated in Figure 3.3, we still need to look at how stellar rotation relates to the potential formation of blue stragglers. Following from our earlier discussion concerning the numerical experiments conducted by Saio and Wheeler [2], rotation will result in a non-zero, non-thermal pressure force term (i.e., $\delta > 0$) coming in to play, and consequentially we might expect rotating stars to have extended main sequence lifetimes. This is certainly to the good, in the sense of making a blue straggler, but it is not the whole story.

Since the centrifugal forces resulting from rotation produce an equatorial bulge, the equations of stellar structure are now somewhat more complicated to solve. In the non-rotating case, the

[6] This law underscores the idea of gyrochronology, in which the age of a low mass star (F, G, K, M spectral types) is estimated according to a measurement of its rotation period. The Sun, in this method, provides an important calibration point since its age and rotation rate are directly measurable: the age, recall, by the study of meteorites, and the rotation period by the study of sunspot motion.

equations can be solved in a one-dimensional sense, with the model properties being described according to the radial distance from the center of the star. In the rotating case, the resulting non-spherical shape requires that the equations of stellar structure need to be solved as a function of both radial distance and angle. The angle term runs from 0 degrees at the equator to 90 degrees at the pole. Mathematically such conditions are well understood, but they do add extra time and complexity to the computational process. For all that, however, the numerical models indicate that during the main sequence phase, a rotating star is initially cooler at its surface (i.e., redder in color) and lower in luminosity than a non-rotating, same-mass comparison star. At the end of the main sequence phase, however, the rotating stars are significantly more luminous than their comparable-mass, non-rotating comparison star. The effect of rotation is to broaden out the main sequence, and it additionally results in longer main sequence lifetimes — these latter effects essentially coming about because the central cores of rotating stars are larger than those found in non-rotating, similar-mass stars.

One of the key physical effects that comes about through rotation is the induction of circulation currents and instabilities within a star's interior. These new physical properties result in the redistribution of material and angular momentum within the star, with the net effect of changing the internal chemical composition profile. The timescale in which the circulation can act to change the composition of a rotating star varies as τ_{KH}/Ω^2, where Ω is the angular velocity $[\Omega(r) = V(r)/r]$ and τ_{KH} is the Kelvin–Helmholtz timescale (recall Equation 2.3). Accordingly, the faster a star rotates, the shorter the circulation timescale, and the more pronounced the redistribution of material (internal mixing) will be. And, as we saw earlier, well-mixed stars, with enhanced non-thermal pressure support, evolve up the main sequence, with higher surface temperature and luminosities into the region where blue stragglers are found in the HR diagram. Accordingly, if fast rotation is a means of producing the internal conditions necessary for the production of a blue straggler, then we might expect to observe that blue stragglers are rapidly rotating stars. The question, therefore, is, is there any evidence for this?

There is not a great quantity of data on blue straggler rotation rates — but there is enough to draw some conclusions. One of the more recent studies is that lead by Alex Billi (Università di Bologna) [4] who, along with colleagues, studied the blue stragglers in the globular cluster NGC 3201. In total, 67 blue stragglers were observed, along with some 114 reference stars. The results showed a distinct difference between the rotation rates for the blue stragglers and the selected post–turn-off stars in the cluster. For the non-blue stragglers observed, none had a rotation velocity greater than 20 km/s. For the blue straggler group, however, a wide spread in velocities up to a maximum recorded of some 200 km/s was observed. Indeed, 28 percent of the blue stragglers studied in the NGC 3201 had 'fast' rotation velocities in excess of 40 km/s. Other studies, such as that by Alessio Mucciarelli (Università di Bologna) and co-workers [5], who looked at some 110 blue stragglers in the globular cluster ω Centauri, also found a large range in rotation velocities, with about 40% of the blue stragglers having velocities greater than 40 km/s, and some 20% with velocities greater than 70 km/s. Similar results for blue straggler spin velocities have been deduced for the globular cluster M 4. The observed rotation velocities of blue stragglers are still well below the critical spin velocity (recall Figure 3.3), but they are indicative of something important going on with respect to blue straggler origins. Interestingly, Billi and co-workers have additionally found that the blue straggler rotation velocity tends to decrease with decreasing luminosity and surface temperature. This suggests that once a blue straggler has formed, in a rapidly rotating state, another mechanism then kicks in to slow the rotation down — in other words, the older the blue straggler, the slower it appears to rotate. In addition to the age-spin-down effect, another distinct trend was also reported in a study of blue stragglers by Francesco Ferraro (Astrophysics and Space Science Observatory Bologna) and co-workers [6]. In their study of 320 high-resolution blue straggler spectra, Ferraro *et al.*, found that the fraction of rapidly rotating blue stragglers (with $V_R > 40$ km/s) decreased with increasing cluster central density. In other words, the mechanism responsible for producing rapidly rotating blue stragglers preferentially works in low star-density environments. In words that Sherlock Holmes might have used, this is a fact to be saved and docketed.

If, as it certainly seems to be the case, rapid rotation and mixing are important mechanisms underlying the production of blue stragglers, we are once again led down the path of another mystery. How do you make a star rotate more rapidly? Is it the case that some stars are simply born as rapid rotators, or is it the case that some additional and/or external component is required? Of these two possibilities, the first concerning naturally born fast rotators is, observationally, unfounded. Stars simply do not appear to form with very high, near critical, rotation rates (recall Figure 3.3). In terms of the second possibility, that some external forcing agent is in play, we are on more productive ground. Indeed, in this latter case, it is the exchange of matter between two stars in a binary system that can result in one star spinning much more rapidly and, in the process, producing a blue straggler. Another possibility is to allow one star to collide with and be subsumed by another star — a situation that will literally turn one star blue and the other into lunch.

3.4. The binary switch

Rush of suns, and roll of systems, and your fiery clash of meteorites
(A. L. Tennyson — *God and the Universe*)

In *The Abominable Bride*[7], Sherlock advises Watson that when it comes to solving a crime, one can always dismiss the idea that there is a twin involved in the subterfuge. The reason (not based upon any specific logic) being, "Because it's never twins." Even the idea of a secret evil twin can apparently be dismissed. In the case of stars, however, it is the twin (evil or otherwise) that often drives the system's story. Indeed, the individual stars within a close binary system do not evolve in isolation but, like conjoined twins, they evolve together, joined not so much at the hip, but at the first Lagrange

[7] A special episode of the TV series *Sherlock* first aired in 2016.

point. Furthermore, as Sherlock also points out in *The Abominable Bride*, "Sometimes to solve a case, one must first solve another."

When it comes to binary star systems, the astronomer is presented with a vast spectrum of new interactions and outcomes to consider. Some twin stars are distant and remote, and the two effectively live separate lives barely aware of their sibling's existence; others are born close and remain close, the one continuously influencing the development of the other. About 1/3 of the stars within the galaxy are found within binary systems. The frequency of binarity, however, depends upon component mass (and hence spectral type). For Sun-like stars (that is stars with a mass in the order of 1 $M_\odot$), the frequency of binarity is about 45%. For low-mass, M-spectral type stars, the binary fraction is about 25%, while for the most massive (O spectral type) stars the binarity fraction is 60% and higher. Accordingly, nature appears to like making single, low-mass stars, and these mostly live out a long and solitary existence. For Sun-like, and more massive stars, the presence of a similar-mass twin is increasingly likely, and if not estranged (due to physical separation), these twins will grow old together, locked in a life-long embrace and a *do-si-do* of orbits. Indeed, the embrace can be sufficiently intimate that the two stars will exchange body-mass, and in some rare cases, two stars will become one. This exchange of mass in binary systems is critical, since as discussed earlier, it is the mass of a star that controls how it will evolve.

3.5. The winking demon

These are much deeper waters than I had thought
(*The Adventure of the Reigate Squires*)

Algol, the second brightest star in the constellation of Perseus, has historically been called the Demon Star. This designation follows from Greek mythology, specifically the battle between Perseus and the demonic Gorgon (Figure 3.4). Algol (or β Persei), the constellated eye of the Gorgon, however, is remarkable for its variability. Indeed, it winks at us every 2.867 days (Figure 3.5). This periodic variability

Figure 3.4. The constellation of Perseus as presented in Johannes Hevelius' *Uranographia* (published in 1690). In this rendition, the left eye of the Gorgon's head reveals the position of Algol (β Persie). The brightest star in the constellation, α Persie (or Mirfak) is positioned close to Persius' nape.

was first described in detail by John Goodricke in 1783. Addressing his thoughts to the assembled Fellows of the Royal Society of London, Goodricke suggested that the dimming of Algol was a consequence of a 'dark star' periodically passing in front of a bright star, thereby partially blocking some of the latter's light in our line of sight. In modern terms, this would make Algol an eclipsing binary system. In his idea, Goodricke was essentially correct, although we now recognize the 'dark', eclipsing companion as being another star. Algol is actually a triple-star system, but the two main (eclipsing binary) components, βPer Aa1 and βPer Aa2, correspond to a B8 V star and K0 IV star respectively. The third star, βPer Ab, is a spectral type F IV star. The stars in the binary component orbit the system's

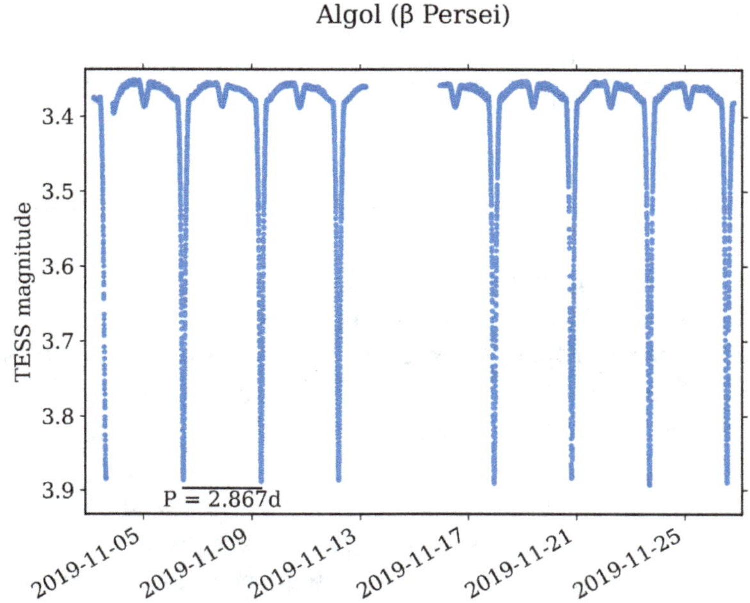

Figure 3.5. Light curve for Algol as recorded by the Transiting Exoplanet Survey Satellite (TESS). The 2.867-day interval between repeated minima is clearly identified. Image courtesy of NASA.

center of mass once every 2.867 days, while the third star orbits the binary system once every 680 days. The binary components βPer Aa1 and βPer Aa2 are remarkably close to each other, being just 0.062 AU apart (this is just 24 times greater than the distance of the Moon from the Earth). The third star, βPer Ab, orbits the binary component at the relatively large distance of 2.7 AU. For all this, it is remarkable to imagine that on the scale of our solar system, the three stars identified in the Algol system could fit in the interior of the main belt asteroid region, with βPer Aa1 and βPer Aa2 easily fitting interior to the orbit of Mercury.

For all of its interesting historical heritage, and the fact that it is a rare eclipsing binary system, Algol is particularly interesting from an astrophysical perspective for the evolutionary status of its component stars. Indeed, they define what became known as the Algol Paradox. Table 3.1 provides a breakdown of the physical properties

Table 3.1. System characteristics deduced for the three stars in the Algol system.

Component	Classification	Mass ($M_\odot$)	Luminosity ($L_\odot$)	Radius ($R_\odot$)
βPer Aa1	B8 V	3.2	182	2.7
βPer Aa2	K0 IV	0.7	6.9	3.5
βPer Ab	F IV	1.8	10.0	1.7

that astronomers have deduced for the stars in Algol. The numbers, as such, are not exceptional in any specific sense — they correspond to those general properties expected for the specific types of stars involved. The paradox lurks entirely within the luminosity class designations that have been deduced for each star in the system (second column of Table 3.1).

Recall from Chapter 1 that the roman numeral following the spectral type designation is an indicator of age. With a luminosity class designation V, βPer Aa1 is a main sequence star, while the other two stars in the system are of luminosity class IV, indicating that they are subgiants. Once again, this is all well and good, and there is no issue with respect to the luminosity class designation. However, and here's the rub, given what we have seen about the speed of evolutionary change associated with stars (specifically the main sequence lifetime), it is the more massive star, βPer Aa1, that should be more advanced — that is, it should have been the giant star instead. It is as if the evolutionary story has been reversed in the Algol stars. Indeed, this is the Algol Paradox. The main sequence lifetime (recall Equation 2.4) of a 3.2 $M_\odot$ star is about 546 million years, while that of a 0.7 $M_\odot$ star is about 24 billion years. Accordingly, βPer Aa2, not βPer Aa1, should be the main sequence component. How is it, therefore, that the masses and luminosity class (that is, evolutionary status) are not in accord with expectation? The answer to this question lies within the process of mass transfer.

In the case of Algol, we are dealing with twin stars, born into a binary system at the same time, with one twin being much more massive than the other. Let us call the two stars A and B, with the initial (formation) masses being such that $M_A > M_B$. Given the mass difference between the stars, we know that the main sequence lifetime of star A is shorter than that of star B — that is, $\tau_{MS}(A) < \tau_{MS}(B)$. Accordingly, star A will evolve off the main sequence and become a giant star long before star B. It is at this stage, however, that because of their close proximity, star A begins to transfer material to star B (the details concerning this will be given below). In this manner, the masses of the twins are rapidly reversed, with star B becoming the more massive of the pair. Accordingly, we eventually end up with a system composed of a low-mass evolved star (star A) in orbit about a more massive, but still, main sequence star (star B). At later times, it is entirely possible that once star B evolves off the main sequence, it will transfer material back to star A. For Algol, while βPer Aa1 is presently the more massive star in the binary, it was initially the least massive component. Conversely, while βPer Aa2 is presently the least massive twin, it was initially the more massive star. There are, in fact, many evolutionary pathways, depending upon the initial masses and separation, that a binary star system can follow, resulting in a whole menagerie of distinct systems coming about — Algol-like binaries being just one such evolutionary outcome (see below).

3.6. The three-body problem

It is quite a three-pipe problem

(*The Red Headed League*)

Gravity is the force that binds; it keeps us from drifting off from the surface of the Earth, and it keeps the Earth in orbit about the Sun. Isaac Newton first showed (in 1687) how the idea of a permanently acting force, directed towards a fixed center, can explain planetary orbits about the Sun. The key features of this (gravitational) force in the solar system is that it varies as the inverse square of the distance from the Sun, and that this, along with the conservation of momentum, will cause the resultant planetary orbits to be elliptical, with the Sun being located at one focus. Furthermore, the Sun-planet line will sweep equal areas in equal amounts of time within the planet's orbit, and the size of a planet's orbit, expressed according to its semi-major axis a, will vary systematically with its orbital period P, with $P^2 \sim a^3$.

Indeed, these conditions of shape, motion, size, and period provided a clear explanation for Kepler's three empirical laws of planetary motion (as first described in 1609 and 1618). In the case of two-body dynamics (e.g., the Sun and a planet, or the Sun and a comet), Kepler's laws are strictly obeyed. In a system of three bodies, or more (e.g., the actual solar system), Kepler's laws are only approximately obeyed, and due to multiple time-varying gravitational interactions between the planets, the orbit of each planet will vary in its shape and spatial orientation as a function of time. On the Earth, such variations result in the Croll–Milankovitch climate variations. The equations of motion representing three-body configurations are notoriously difficult to solve for, but if simplifications are introduced, then a sensible picture of the system can be built up. The simplifications that are usually introduced are that the two stars in the binary system have circular orbits about their common barycenter, and that the mass of the third, smaller body is very much less than that of the stars (and can be effectively set to zero). Under these conditions the contours of constant effective gravitational potential, Φ, can be mapped out, and the dynamics of the smaller body can be readily described. The term effective is used since the centrifugal potential due to the system's rotation must be included in the calculation. In a 2-dimensional system, the effective gravitational potential will be

Φ = gravitational potential of star 1 + gravitational potential of star 2 + centrifugal potential

Letting the two stars be located at positions (0, 0) and (*a*, 0) respectively, with *a* being the separation between their centers, in a binary system with angular rotation velocity Ω, the effective gravitation potential at some arbitrary point (X, Y) will be

$$\Phi = -\frac{GM_1}{\sqrt{X^2+Y^2}} - \frac{GM_2}{\sqrt{(X-a)^2+Y^2}} - \frac{\Omega^2}{2}[(X-\mu a)^2+Y^2] \qquad (3.5)$$

where μ is the reduced mass, $\mu = M_1\, M_2/(M_1 + M_2)$. This equation is rather complex, but allows us (with the aid of a computer) to map out lines of constant potential. Figure 3.6 reveals a set of equipotential contours, for Φ = constant, in a binary system composed of two stars, with masses M_1 and $M_2 < M_1$. Closer to each star, the contours (not actually shown) will be approximately circular. On the line drawn between the star centers there will be a critical point, the so-called first Lagrange point (L_1), at which a test-particle will dither

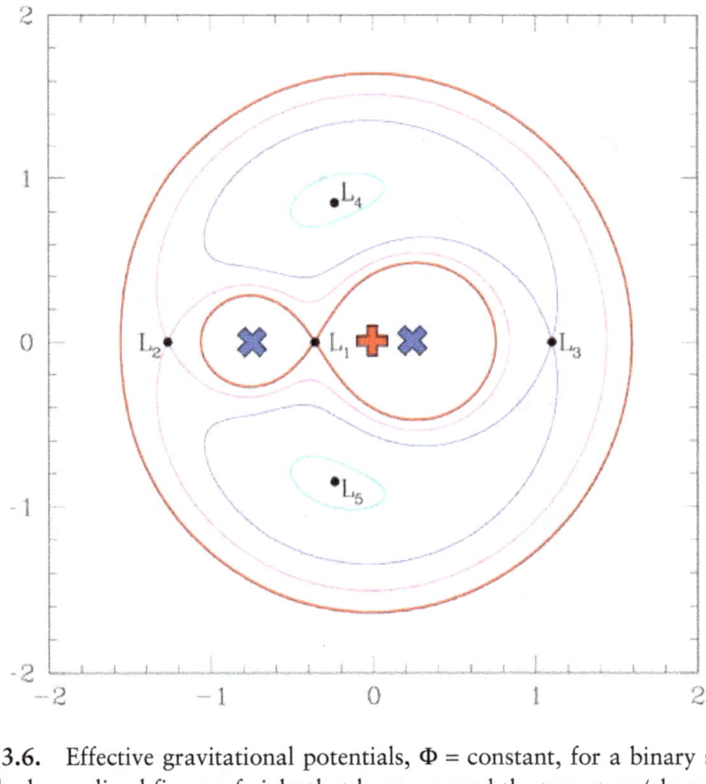

Figure 3.6. Effective gravitational potentials, Φ = constant, for a binary star system. The heavy-lined figure-of-eight that loops around the two stars (shown by the blue × symbols) corresponds to the Roche lobes associated with each star. Five Lagrange points are indicated in the diagram, with L_1 being the location where mass can be transferred from one star to the other. From L_2 and L_3, matter can leak-out of the system. L_4 and L_5 correspond to stable, so-called, Trojan points. At large distances from the two central stars, the potential approaches that of a single object having the same mass as the two components (large outer red circle). The center of rotation for the system (marked by a red + symbol) is located at the origin (0, 0). Image adapted from Henri Boffin [7].

on the knife-edge of instability, with just the smallest of nudges carrying it towards one star or the other. The critical equipotential contour that passes through the L_1 point is called the Roche lobe, after French mathematician Édouard Roche who studied such systems in the mid- to late-1800s. The Roche lobe traces out a figure eight (double teardrop) contour about the two stars, and is shown in red in Figure 3.6. Material located within the Roche lobe about a given star will remain gravitationally bound to that star. Material located outside of the Roche lobe is more mobile, and may be accreted by one of the stars, or even be lost from the system through the L_2 or L_3 points. At large distances away from the binary configuration, the equipotential contours become circular, and the motion is equivalent to that about a single point mass having a mass equal to that of the two system stars.

Where all this is of relevance to our story concerning the Algol paradox is that if one star has swollen sufficiently in size (as a result of its age and evolution) to approximately fill its Roche lobe, then material from the bloated star can be transferred through the L_1 point to its companion — this is generally called Roche lobe overflow (RLOF). Literally, in the case of RLOF, the secondary star will gain mass at the cost of the primary losing it. The precise shape of the Roche lobe depends upon the mass ratio of the two stars within the binary system: $q = M_1/M_2$ (in Figure 3.6, $q = 3$). Peter Eggleton (Institute of Astronomy, University of Cambridge [8]) has published a number of simplified approximation formulae for the Roche lobe radius R_{RL} and finds that in the range $0.1 < q < 10$, a good approximation is:

$$R_{RL} = \frac{0.44\,a\,q^{0.33}}{(1+q)^{0.2}} \tag{3.6}$$

where a is the orbital separation of the two stars (recall Chapter 1.7). Accordingly, the greater the mass ratio q, and the greater the separation a, the larger the Roche lobe about each star. It can be shown from Equation (3.6) that the ratio of the Roche lobe radii will vary as $R_{RL,1}/R_{RL,2} = q^{0.46}$. For $q = 3$, as in Figure 3.6, this indicates that the approximate Roche lobe radius of the secondary (M_2) is some 60% smaller than the Roche lobe radius of the primary (M_1).

Mass transfer from the primary to the secondary (that is, RLOF) will occur once the radius of the primary star has expanded to a level where $R = R_{RL,1}$. Note, the size of the Roche lobe changes as the system evolves, since M_1 is getting smaller and M_2 is getting larger, so the ratio q will be decreasing; the mass exchange will also reduce the orbital separation a between the two stars (see next section).

3.7. Mass exchange

It is a tangled skein, you understand, and I am looking for a loose end

(*The Adventure of the Creeping Man*)

When binary stars interact, if the conditions are right, mass flows. The conditions of interaction are shown schematically in Figure 3.7. The key requirements for mass transfer to occur are that we have a binary system in which one star is more massive than the other, and the two stars are relatively close to each other. The more massive star will evolve off the main sequence and become a giant more rapidly than its lower mass companion. Eventually, the radius of the more massive, primary star begins to fill its Roche lobe while the lower-mass secondary is still in its main sequence phase [corresponding to Figure 3.7 (b) and (c)]. Once the Roche lobe of the primary is filled, that is, $R_1 = R_{RL}$, matter can begin to leak through the L_1 point to be accreted onto the secondary star. We now have a situation where the primary star is losing mass, while the secondary is gaining it. This mass exchange will begin to drive a number of system changes — changes that will alter the large-scale properties of the binary system, and changes that will alter the very course of the canonical evolution normally associated with each individual star.

Once mass transfer begins to occur, a large number of possible evolutionary outcomes become possible — these outcomes being entirely dependent upon the initial masses and the initial orbital separation of the two stars. Key to unraveling what a given system

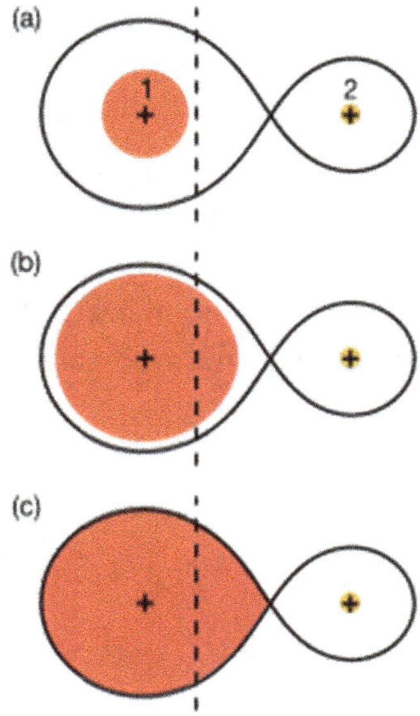

Figure 3.7. Stages in the life of a close binary system. The dashed line is the system rotation axis, and initially the mass of star 1 (on the left) is greater than the mass of star 2 (on the right). (a) Both stars lie within their respective Roche lobes and they effectively evolve independently of each other. (b) Star 1, the initially more massive star, has grown to nearly fill its Roche lobe, its evolutionary timescale being more rapid than that of star 2. (c) Roche Lobe overflow stage: Star 1 now fills its Roche lobe and will begin to transfer material to star 2 through the L_1 point.

will do, however, is the exchange of angular momentum. Importantly, it is conserved. That is, it is taken to remain constant at every instant, even though the masses of the two stars are changing. We can proceed by looking back to Chapter 1, and especially Section 1.7. Equations (1.14) and (1.15) give us relationships between the masses of the primary and secondary stars (M_1 and M_2), and their respective distances a_1 and a_2 from the center of mass (we assume circular orbits for simplicity here), where $a_1 + a_2 = a$, and a is related to the sum of the star masses and the orbital period P through Kepler's 3rd law. The stars move along their orbital paths at a speed $V_1 = 2\pi\, a_1/P$ and $V_2 = 2\pi\, a_2/P$. Additionally, these velocities are related to the angular velocity Ω through the relationships: $V_1 = a_1\, \Omega$, and $V_2 = a_2\, \Omega$. Now the angular momentum of the primary is $J_1 = M_1\, a_1\, V_1$, and that of the secondary is $J_2 = M_2\, a_2\, V_2$. The total angular momentum of the binary is therefore $J = J_1 + J_2 = M_1\, M_2\, a^2\, \Omega/(M_1 + M_2)$. The last equality just shown follows from an algebra step, but the key point is that J is to remain constant even when the mass of the two stars begins to change. We take it that the $(M_1 + M_2)$ term, the system mass, remains constant — this is not always true, and some material will inevitably leak from the system during the mass transfer stage. Accordingly, we have an interplay between the product of the individual masses and the orbital separation of stars, a, to play with in order to keep J constant. For example, if we have a system with two stars of initial mass 2 $M_\odot$ and 1 $M_\odot$, with an initial orbital separation of $a_{\text{initial}} = 10$ AU, $J_{\text{initial}} = 2 \times 1 \times 10^2 \times K$, where $K = \Omega/(M_1 + M_2)$ is a constant. If, say, the primary transfers 0.5 $M_\odot$ to the secondary, then $J_{\text{final}} = 1.5 \times 1.5\, a^2_{\text{final}}\, K = J_{\text{initial}}$, and accordingly $a_{\text{final}} = 9.429$ AU. In other words, the system is more compact, with the two stars becoming closer by 0.571 AU. This will be observable, through Kepler's 3rd law, as a change in the observed orbital period. The change in the orbital period will be $\Delta P = 3\ (P/a)\ \Delta\alpha/2$, where $\Delta\alpha = a_{\text{initial}} - a_{\text{final}}$, and P and a are the initial orbital values (18.26 years and 10 AU respectively). Accordingly, the orbital period decreases by $\Delta P = 1.56$ years to $P_{\text{final}} = 16.70$ years. As long as mass is being exchanged from one star to the other, the system's orbital parameters will change. The size of the orbit will increase (along with the system's orbital period) once the mass of the (initial) primary is smaller than that of the (initial) secondary. In addition, the size of the Roche lobe appropriate to the primary will vary according to the changes in q and a, as indicated in Equation (3.6).

How an individual star reacts to losing or gaining mass will depend upon how rapidly its mass changes and upon its evolutionary status. Three cases of RLOF are generally recognized: cases A, B and C. If RLOF occurs when the donor (initial primary) star is still in its main sequence phase, then case A overflow is said to occur. If RLOF occurs prior to the onset of helium ignition in the donor star, then case B overflow is said to occur. Case C overflow is said to occur if the donor star has initiated helium fusion reactions within its core.

How a star behaves to losing or gaining mass will depend upon a number of factors. The mass transfer timescale τ_{trans} can be defined according to the donor's mass such that $\tau_{trans} = M_1/M_{trans}$, where M_{trans} is the mass transfer rate (typically expressed in $M_\odot$/yr). Now, the star gaining mass will be out of thermal equilibrium if $\tau_{trans} < \tau_{KH}$, the Kelvin-Helmholtz timescale (recall Equation 2.3). In this state, the star will expand until thermal equilibrium is regained. This expansion may cause the star gaining mass to fill its Roche lobe, resulting in the formation of a contact binary (see Section 3.10). If $\tau_{trans} > \tau_{KH}$, then the gainer will evolve on a nuclear timescale τ_{nuc} (recall Equation 2.4). During mass transfer the radius of the donor star can either shrink or expand. Mass transfer is stable provided we have the condition that $R_1 \approx R_{RL}$ at all times. We have seen, however, that as the mass transfer proceeds so the separation between the two stars decreases, and this reduces the size of R_{RL} (recall Equation 3.6). If the donor star cannot shrink faster (as a result of its mass loss) than the size of its Roche lobe, then it will increasingly overfill its lobe, resulting in an even higher mass transfer rate. This increased mass transfer rate will cause the stars to move even closer together, further reducing the size of the donor star's Roche lobe, and a runaway, rapid mass transfer condition will result. Indeed, once this runaway condition sets in, matter is transferred on a near dynamical timescale, with the donor star eventually losing so much mass to its companion that it becomes the less massive component.

3.8. A model for Algol

I am becoming in coming undone. I unbind.

(Sarah Gerard — *Binary Star*)

We are now in a position to understand the Algol system and say something about its early history. Figure 3.8 shows the evolutionary tracks for a binary system in which the initial masses were taken as 3.4 $M_\odot$ and 1.1 $M_\odot$, and the orbital period was 1.4625 days [9]. The evolutionary track of the initially more massive star is shown in red; that of the initially lower mass star in blue. We know, from earlier

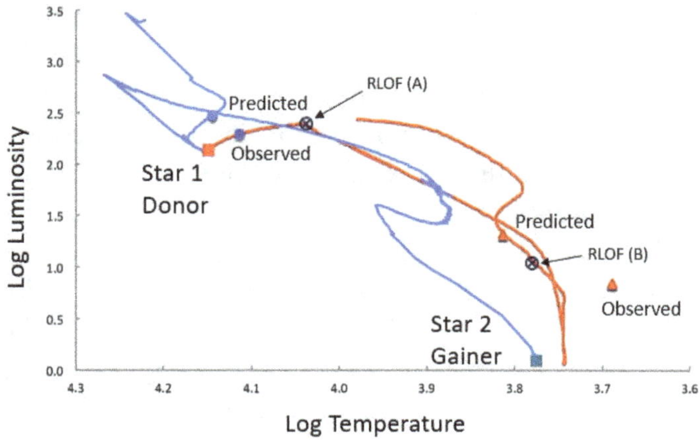

Figure 3.8. Evolutionary tracks for Algol (β Persei) in the HR diagram. The initial main sequence location of each star is indicated by a square (red for the donor star, blue for the gainer). The observed and predicted model calculations, for a system having an age of 310 million years, are shown by blue circles for the gainer, and red triangles for the donor. The locations at which the donor star undergoes Roche lobe overflow (A) and (B) are indicated by a ⊗ symbol. The computations have been run up to the time at which the donor star initiates core helium-burning (at a system age of about 361 million years). Image adapted from [9], and model data kindly supplied by W. van Rensbergen (Universiteit Brussel).

discussion, that the more massive star is going to evolve more rapidly than its lower mass companion; the primary having a main sequence lifetime of about 466 million years, versus 7.9 billion years for the secondary. Indeed, we see in Figure 3.8 that before the lower mass (gainer) star moves very far from its initial location on the main sequence, the more massive (donor) star has evolved right across the main sequence band, increasing in size and luminosity, until it fills its Roche lobe when the system is about 210 million years old [RLOF (A) in Figure 3.8]. Once RLOF begins, mass is transferred from the donor to the gainer star, at rate of about 10^{-9} $M_{\odot}$/yr, with the effect that the evolutionary track of the donor star moves to much lower luminosities and effectively down the main sequence (recall Figure 2.3). The gainer, in contrast, evolves upwards along the main sequence to higher luminosities and temperatures. As time ticks by, the primary continues to lose material, through the L_1 location (recall Figure 3.6), to be accreted by the secondary. After a system age of about 254 million years, the donor star exhausts hydrogen in its core, and prior to the ignition of core helium-burning, enters its RLOF (B) phase. Mass continues to be exchanged between the stars until the masses are essentially reversed from their original values (recall Table 3.1) — the gainer becomes the more massive star, in fact, when the system is some 212.6 million years old. When the system is about 310 million years old, and the donor star has lost about 2.6 solar masses to its companion, the gainer is effectively a young 3.7 $M_\odot$ main sequence star. The system orbital period has now also changed from its initial 1.4 days to some 2.8 days.

Having presented all the above with respect to Algol, the question now becomes, how does all this help us with respect to the blue straggler mystery? Indeed, is there any direct observational evidence to support of the idea that at least some blue stragglers have been produced as a result of mass transfer? To answer this question in the affirmative, we direct our attention to the 7-billion-year-old, open cluster NGC 188. The observations indicate that there are some 20 blue stragglers in the cluster, and radial velocity variations in the spectral line characteristics of these stars indicates that at least 16 (that is 80%) of them are located within close binary systems. Of the remaining four blue stragglers, three show no apparent radial velocity variations in their spectral lines, and one is apparently spinning so rapidly (resulting in very broad spectral line features), that no clear radial velocity variations can be detected. In addition to the high binary percentage among the blue stragglers in NGC 188, spectral observations obtained with the Hubble Space Telescope, by Natalie Gosnell (Colorado College) and co-workers [10], find that three of the blue stragglers have hot and young white dwarf companions. The presence of the white dwarf companions is betrayed by an excess of ultraviolet radiation in the spectra of the blue straggler stars — that is, there is more energy being radiated in the UV part of the spectrum than would be expected from the blue straggler on its own. Furthermore, by modeling the UV excess in the blue straggler spectra, Gosnell *et al.* found that the white dwarfs in the systems that they observed are no more than 300 million years old, indicating that mass transfer has only recently ended for these particular systems.

3.9. Beyond Algol — the McCrea hypothesis

The Phenomenon has caused considerable perplexity
(W.H. McCrea, *Monthly Notices RAS*, **128**, 1964)

The first detailed investigation of the binary mass-transfer mechanism for making blue stragglers was conducted by William McCrea in a research article, published in the journal of the Royal Astronomical Society, in 1964 [11]. Entitled *Extended main-sequence of some stellar clusters*, McCrea set out to study what the consequences of mass transfer between binary stars might be, and especially how the evolutionary consequences might play out in "cluster HR diagrams". Indeed, he specifically singled out the observation that some clusters show an extension of the main sequence that are brighter than the canonical cluster turn-off. The "blue stars" of interest to McCrea are our blue stragglers. He noted that such an extension does not comply with the accepted idea that cluster stars all formed at the same time from the same initial interstellar cloud, and he asked if they might have "formed in some separate process long after the formation of the cluster as a whole." In answer to this, however, he cites his then-doctoral student's, Iwan Williams', work and argues that time delay in the formation process was not the answer to the blue straggler problem; neither, for that matter, is accretion of material from the interstellar medium (as we discussed above). Rather, McCrea argued that the Algol phenomenon provides the key to answering the question: why are there blue stragglers? In the introduction to his research paper, McCrea sets down the following hypothesis, "the apparently young stars [i.e., blue stragglers] are secondaries in close binary systems to which mass has been transferred from the original primaries." After this, his aim was to "see whether the hypothesis predicts the existence of 'blue stars' in the types of clusters in which they are found." He further wanted to determine what it is that might be learned "about stellar clusters by [the] further study of the 'blue star' phenomenon."

The first result that McCrea establishes is made on the basis that blue stragglers are formed by mass transfer within binary systems. Accordingly, the maximum mass that any blue straggler within a cluster might conceivable have is 2 M_{TO}, where, recall, M_{TO} is the turn-off mass appropriate to the cluster. Using the luminosity-mass law determined for main sequence stars (recall Figure 1.13), this result tells us that, with respect to their luminosity, $L_{max,BS} \approx 2^{\alpha} L_{TO}$, where $\alpha \approx 3.5$ is the index appropriate to the luminosity-mass law for near solar mass stars. Now, since the blue stragglers are cluster members, located at the same distance from the Earth as the other cluster stars, luminosities can be substituted for fluxes in Equation (1.7), and this allows for a maximum magnitude difference Δm between that of the brightest possible blue straggler and that of the turn-off magnitude for the cluster to be determined. Accordingly, $\Delta m < 2.5 \log (2^{3.5}) = 2.6$ mag, and this result is borne out by the observational data. No blue stragglers more than 2.6 magnitudes brighter than the cluster turn-off magnitude have ever been observed.

McCrea's full theoretical argument need not be considered here, but he shows that the idea of mass transfer between stars within a close binary system is consistent with producing blue stragglers, and he draws out several observational "consequences":

1. A cluster that has not produced any red giants cannot have any blue stragglers.
2. A cluster that has started to produce red giants has a "blue extension" if and only if, some of the stars that would have become giants possess close companions that were initially of smaller mass.
3. Blue stragglers can be no more than about 2.6 magnitudes brighter than the magnitude corresponding to the cluster turn-off.
4. All blue stragglers must be located within a binary system.
5. There should be a small number of stars (effectively evolved blue stragglers) lying between the main sequence and the giant branch.

To these "consequences" McCrea further adds, "if the ideas are correct, it would be surprising if any sufficiently old galactic cluster does not show the phenomenon [of blue stragglers]." Indeed, consequence one is borne out by the observations, and, for example, the very young Pleiades open cluster, with an age of some 75 to 150 million years, shows virtually no giants in its HR diagram and has no blue stragglers. This consequence is also supported by the data plotted in Figure 2.7, where it is evident that clusters younger than a few hundred million years old do not contain any blue stragglers. Consequence two is also partially verified, in the sense that binary star systems are certainly observed within star clusters. Consequence five, as we shall see below, has also been borne out by the observations. As testament to McCrea's foresight and analysis, it is only consequence four that needs present day modification. McCrea did not consider the idea of direct star-on-star collisions, nor the possibility of binary merger events (see below). McCrea's 1946 analysis did not include any predictions based upon the numerical models of stellar evolution — indeed, it could not, since such model results were still some 30 years away from being computationally possible. For all that, he concluded his paper with a discussion of where such calculations, once available, might take the story, noting that, it will be of "interest to pursue the full history of what we have been calling blue stars." McCrea was without doubt right on this count, although the riddle of the "blue stars" was far from being fully settled then, as indeed, it still is now.

3.10. When two becomes one

Figure 3.7 illustrates the situation of close binary system evolution up to the point of forming a semi-detached binary. That is, Figure 3.7(c) shows the situation where one star, the initially more massive star, has filled its Roche lobe and the process of mass transfer has begun. This need not be the final stage of evolution, however, and it can be the case that both stars evolve to a state in which they both fill their respective Roche lobes. In this situation a contact binary can form, and this may further result in the formation of a common envelope phase. These latter phases are shown schematically in Figure 3.9. The distinctive, rounded and double-humped light curve variations observed from so-called W Ursae Majoris stars are explained through the idea that they are actually contact binaries, in which both stars fill their respective Roche lobes (Figure 3.9(d)). Importantly, during the contact binary phase, the common gas envelope rotates at the same rate as the stars within the binary system. In contrast, however, during a common envelope phase (Figure 3.9(e)), the envelope rotates at a slightly slower rate than that of the stars within the binary system. This latter rotation difference is critical, since it sets up drag forces on the two stars in the binary system, causing the distance between them to decrease. Indeed, if the common envelope is not rapidly ejected from the system, then the two stars will ultimately merge.

Depending upon the initial system parameters, the contact and common envelope phases are thought to be responsible for

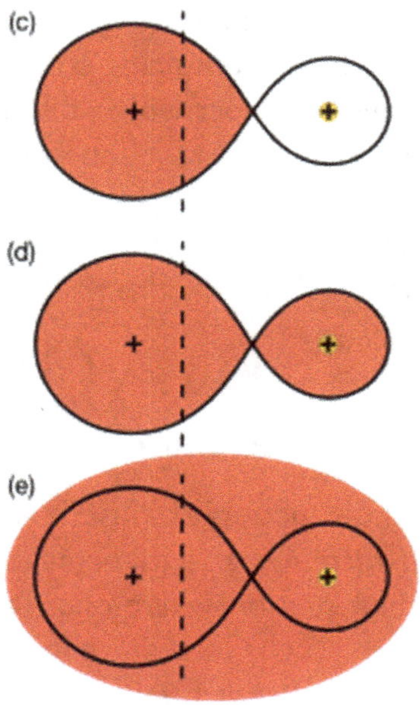

Figure 3.9. The continuation of mass exchange between two stars within a binary system, leading from a contact system (c) [recall also Figure 3.7], to a contact binary (d) and a common envelope phase (e).

producing so-called cataclysmic variables (in which a white dwarf or neutron star and a mass-transferring companion are in very close proximity) and systems composed of two close white dwarfs, and even, on occasion, two close neutron stars. The occurrence of a common envelope phase and mass transfer within a binary system has also been linked to the appearance of several 'exotic' astrophysical phenomena. In cataclysmic binaries we have a degenerate star (a white dwarf or a neutron star) accreting matter from a Roche lobe-filling companion. The matter lost from the donor in this case can either accrete directly onto the surface of the gainer, or, if the gainer supports a strong magnetic field, then it will accrete primarily onto its (magnetic) polar regions. Alternatively, an accretion disc can form around the gainer, with a hot spot appearing where the matter transferred from the donor feeds into the disk. Furthermore, mass transfer can act to increase the spin rate of the donor star, and merger events will not only mix the accreting star's interior, it may also significantly increase its magnetic field and alter its surface chemical composition. Astronomers have found evidence for all of these various phenomena taking place. It is also the case that some blue stragglers have been found with anomalously high spin rates, and other blue stragglers have been found supporting accretion disks. These observations all feed into the idea that one of the most important channels for blue straggler formation (recall Figure 3.1) is that related to close binary star evolution, and specifically, stellar mergers.

3.11. The story of V838 Monocerotis

This incident gives zest to our investigation

(*The Speckled Band*)

Theoretical scenarios of mass exchange and mergers are all well and good, but is there any direct evidence for stellar cannibalism? The answer to this question is yes. One situation where this may indeed be the case is that exemplified by V838 Monocerotis in early 2002. It was literally a flash in the night that caught the eye of astronomers. On the night of January 6, 2002, Nicholas Brown, at Quinns Rocks in Australia, thought he had discovered a nova. He found on his exposed film a faint, magnitude 10 star in the constellation of Monoceros, where no star had been recorded in earlier exposures. The sudden brightening of V838 Monocerotis had all the hallmarks of being a nova, but continued monitoring, over many days, weeks and months indicated that something entirely different was happening. Indeed, the brightness variations of V838 Monocerotis turned out to be unique. Figure 3.10 shows the variation in brightness of V838 Mon., as recorded by the American Association of Variable Star Observers (AAVSO), between January and May of 2002. Remarkably, the initial brightening caught by Nicholas Brown on January 6th, was just the prelude to what followed. Indeed, a much more dramatic brightening occurred in early February, when briefly, on February 6th, V838 Mon. peaked at a visual magnitude of 6.75 (not quite, but nearly making it visible to the naked eye). There then followed another decline in brightness, only for another increase to occur in early April. All in all, there were three distinct peaks in the light curve of V838 Mon. and this is something that novae and supernovae just don't do. The second peak is particularly dramatic for what it implies. At a distance of 6,000 pc, V838 Mon. reached

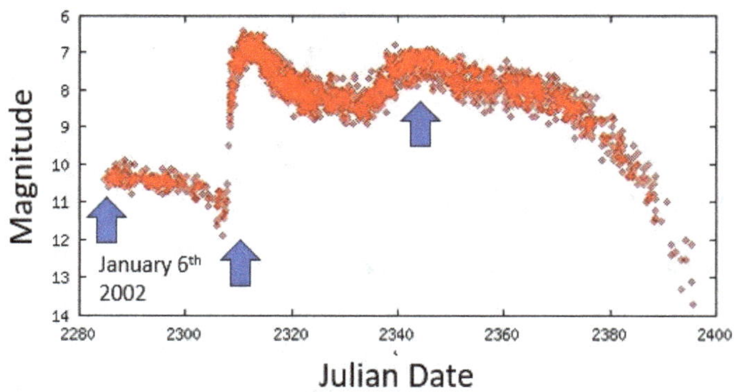

Figure 3.10. Light curve development of V838 Monocerotis. Brightness variations are shown over a 120-day timespan from 5 January to 5 May, 2002. The three arrows indicate the occurrence times of three distinctive peaks in magnitude — see text for details. Image based upon data gathered by AAVSO.

an absolute visual magnitude (recall Equation 1.9) of $M_V = -7.1$, and this equates to its luminosity being some one million times greater than that of the Sun. Not only this, estimates for the size of V838 Mon. indicated that it was some 1,500 times larger than the Sun. At its peak in February, V838 Mon. was one of the largest and brightest stars in the entire galaxy.

Not only was the outburst of V838 Mon. remarkable, so too was what appeared around it. The burst of light that was emitted during the sudden brightening of the system, increasingly moving into the surrounding space, illuminated a vast assembly of surrounding dust clouds (Figure 3.11). What we are seeing in this case is light from the

Figure 3.11. Hubble Space Telescope images showing the development of the light echo associated with V838 Monocerotis. The inset images, to the left, show the development of the light echo at various dates between 20 May 2002 (top) and 8 February 2004 (bottom). Image courtesy of NASA.

outburst being reflected back into our line of sight, and accordingly, we are seeing the development of a light echo. The inset images to the left in Figure 3.11 were captured by the Hubble Space Telescope, and show the development of the light echo from May 2002 to February 2004. It is not clear if the dust revealed in the light echo is physically associated with V838 Mon., or just a natural component of the star formation region in which the system sits. The characteristic red color associated with V838 Mon. and its slow decay from maximum brightness resulted in V838 Mon. being the essential prototype for what are now known as Luminous Red Novae (LRN). Very few LRN have ever been observed, but another example was recorded in September of 2008, when V1309 Scorpio suddenly brightened in a manner similar to that observed in V838 Mon.

A number of interpretations of what happened to produce the outburst of V838 Mon. are available, but the consensus model is that relating to a stellar merger. In this case the two stars within a compact binary system literally melded and combined to become a single star. Noam Soker (Technion, Israel) and co-workers [12] developed this idea in a 2006 research paper, and argued that the outburst was the result of an 8 $M_{\odot}$ (B-spectral type main sequence star) absorbing (that is, stripping and merging with) a 0.3 $M_{\odot}$ pre-main sequence companion star. In contrast to the stellar merger hypothesis, Alon Retter and co-workers [13] suggested in a 2006 research paper that the brightness variations associated with the light curve of V838 Mon. are suggestive of three merger events (as arrowed in Figure 3.10) having occurred. What was actually happening, Retter and co-workers argue [13], was not the merger of two stars, but the accretion and destruction of three large planets in the outer layers of their parent star. In this case, the planets spiraled in towards the central star and were rapidly ripped apart and accreted, one by one, over a time span of about 60 days. The observations surrounding the appearance of V1309 Scorpio have additionally been interpreted in terms of the merging star scenario. Indeed, Romuald Tylenda (Copernican Astronomical Centre, Poland) and co-workers [14] suggested in 2011 that this particular event was related to a 1.5 $M_{\odot}$ star merging with and then absorbing its 0.16 $M_{\odot}$ companion. Although, in similar vein to V838 Mon., the light curve for V1309 Scorpio shows two distinct peaks, and accordingly, a planetary accretion model might also explain the observations. Definitive evidence for a planetary accretion event has been found in the system ZTF SLRN-2020, first recorded by the Zwicky Transient Facility on 20 May 2020. A detailed study of this brightening event, which lasted some 6 months, was published by Kishalay De (Columbia University) and coworkers in the 3 May 2023 issue of the journal *Nature*. In their study, De *et al.* found that the observed brightening was the result of a near solar mass (K spectral type) star gradually ripping apart and then subsuming the material from a planet estimated to be about 10 times more massive than Jupiter. Given, however, the ordinary main sequence characteristics of the star that brightened in this particular outburst event, De and co-workers conclude that the accretion and brightening was a result of the planet's orbital decay, rather than a consequence of the host star expanding and enveloping the planet.

3.12. The story of the Dragon's Egg

The two events are connected — must be connected. It is for us to find the connection

(*The Second Stain*)

The bipolar emission nebula NGC 6164 is a visual feast for the eye, and a marvel to behold (Figure 3.12). Dubbed the Dragon's Egg Nebula, what we are seeing in NGC 6164 is a vast cloud of glowing gas and illuminated dust. At the core of the Egg is a binary system, cataloged as HD 148937, composed of two massive stars. Located some 1,200 pc away, the nebula is large, being some 1.3 pc across, and it was discovered in the southern constellation of Norma by John Herschel in 1834. It has long been an object of interest to astronomers. Indeed, one of the stars in HD 148937 is classified as a very rare Of?p spectral type star. These stars are massive, highly luminous and the f?p qualifier indicates that it shows a peculiar spectrum with emission lines due to ionized nitrogen and carbon. This latter observational feature indicates, in fact, that the star has a hot, extended atmosphere with a high mass loss rate. Not only this, one

Figure 3.12. The Dragon's Egg Nebula, NGC 6164, associated with the binary star system (at image center) HD 148937. Image courtesy of ESO.

of the intriguing peculiarities of HD 148937 is that it supports a magnetic field. Among massive stars, such magnetic fields are relatively rare (recall Section 3.3.2 above). Further to all these peculiarities, HD 148937 is also the only galactic Of?p star to be surrounded by a nebula. Peculiar and unique: clearly something interesting is going on at the core of the Dragon's Egg.

Most massive stars are found in binary or higher order multiple systems, and in the case of HD 148937 the system consists of a binary with a 29.9 $M_\odot$ primary and a 26.6 $M_\odot$ secondary, with the system having an orbital period is 25.76 years. Since the primary (by definition) is the more massive star, canonical wisdom dictates that it should have advanced more rapidly along its evolutionary track than the lower mass secondary. Upon checking this standard stellar evolutionary wisdom, however, Abigail Frost (European Southern Observatory) and co-workers [15] were surprised to find that the more massive primary in HD 148937 was younger (in a main sequence evolutionary sense) than the less massive secondary. Comparing model calculations against the observed luminosity and temperature of the primary indicated an age of 2.68 million years. The deduced properties for the secondary, however, suggest an age of between 4.1 and 6.58 million years. The primary is, apparently, a few million years younger than its secondary companion. Since the two stars are in a binary system, they must have formed at the same time — accordingly, we have a mass/age mismatch. The key observational feature that points toward an explanation for the apparent youth of the primary, according to Frost and co-workers, is that the primary supports a magnetic field. This property of supporting a magnetic field is, recall, unusual because massive stars do not have an outer convective layer to sustain a long-lived (dynamo-driven) magnetic field. So, the question is, why does the primary, in spite of its mass, support a magnetic field? The answer to this lies in the understanding that the magnetic field we observe today is not the remnant of a primordial field, but is in fact just a few thousands of years old. Frost and co-workers argue that, in fact, the magnetic field was actually generated by a merger event. Indeed, the merger event not only stirred-up and mixed the outer layers of the primary star, enabling a temporary magnetic field to form; it also ejected the material that we now see as the Dragon's Egg Nebula. Furthermore, we can also estimate when the merger occurred since the nebula has an estimated age of some 3 to 7.5 thousand years. The evolutionary history for HD 148937 is now revealed. The system, it transpires, formed some 4.1 to 6.58 million years ago as a triple star system. The numerical simulations suggest that the triple system was initially composed of a secondary, with a mass of some 27 $M_{\odot}$, and a primary made-up of a compact binary, with the compact binary being composed of a 5 $M_{\odot}$ star in orbit about a 30 $M_{\odot}$ star. As the system aged, the primary binary system gradually evolved to undergo a merger event between 3 to 7.5 thousand years ago. At this time, the material that is now seen in the Dragon's Egg Nebula was ejected from the system, the circumstances for generating a magnetic field were enabled in the outer layers of the 30 $M_{\odot}$ star, and as a result of merger-induced internal mixing, the evolutionary clock of the 30 $M_{\odot}$ star (now the sole primary star) was set back by a million or more years. As the system continues to age, however, the combined effects of the magnetic field and mass loss from the primary will act to slow its rotation rate. And finally, after about a million years or so, the magnetic field on the primary will have decayed away, and what will appear is a binary system in which the primary appears younger than its companion secondary — we have now, in fact, a massive blue straggler located within a binary system.

3.13. Making *Baekdu*

We all have a past, Watson. Ghosts.

(*Sherlock* — TV series)

The case of *Baekdu* provides us with another example of a possible binary merger producing a blue straggler, but this time with a Jovian planet companion (called *Halla*[8]). The star *Baekdu* (also designated 8 Ursae Minoris) is spectroscopically classified as an G III subgiant, and has a mass about 1.5 to 1.7 times that of the Sun. Discovered by Brian Lee (Bohyunsan Optical Astronomy Observatory, South Korea) and co-workers [16] in 2015, the Jovian planet (*Halla*) occupies a close-in orbit of just 0.46 AU. This is an intriguing result in the sense that the planet could not have formed at the orbital distance that it now occupies, and, furthermore, if it was at its present orbital distance when the parent star (*Baekdu*) was in its red giant

[8] *Baekdu* and *Halla* are the names of the tallest mountains in North and South Korea: Baekdu and Hallasan respectively.

phase, then it should have been consumed in the stars outer envelope. Clearly, something interesting has happened and Lee *et al.* suggest that perhaps the film *Star Wars*[9] might provide an answer to the apparent enigma. Indeed, they suggest that what we now see as the single star *Baekdu* was originally a close binary pair that merged (Figure 3.13). The planet *Halla*, however, survived because it was in

Figure 3.13. Artist's impression of the merger phase of 8 UMi (*Baekdu*). The Jovian planet (*Halla*) is shown to the lower left, along with potential debris from an asteroid belt, or possible material ejected from a disrupted terrestrial planet. Image credit: W. M. Keck Observatory/Adam Makarenko.

[9]Technically, it is *Star Wars episode V — A New Hope* and the home planet of Luke Skywalker and Tatooine, which occupies an external orbit about a binary star system composed of Tatoo I and Tatoo II, that is being invoked.

an external orbit about the binary, rather than in orbit about one of the stars in the initial system. As the stars in the binary merged, producing a blue straggler (that is *Baekdu*), *Halla* then migrated inwards to its present orbital radius. That we are dealing with a merger event and the production of a blue straggler is made clear in light of observations conducted by Huiling Chen (Peking University) and co-workers [17] in 2024. Obtaining photometric observations of *Baekdu*, they applied isochrone fitting methods (recall Chapter 2) to determine the system's age to be 1.9 billion years. This is problematic since the mass of *Baekdu* (about 1.5 solar masses) would suggest that its main sequence lifetime should be at least 3 billion years. The age discrepancy, however, is a result of the evolutionary clock of *Baekdu* being reset during the binary merger process.

3.14. Star collisions

Chaos, cosmos! Cosmos, chaos! Who can tell how all will end
(A. L. Tennyson — *Locksley Hall sixty years after*)

The story of HD 148937 sets the scene for what is potentially an efficient way of creating a blue straggler — it is the end result of companion cannibalism within a close binary system. The elixir of life for such a star, that is, the means by which it can reset its evolutionary clock, is that of gaining mass, and in the process, partially mixing up its interior. The would-be blue straggler forms, as it were, by literally consuming its erstwhile sibling. This is possible, of course, if the star being rejuvenated is born into a close binary system with a just right initial mass ratio and orbital separation. So, we can form blue stragglers in binary systems, but what about single stars? Are they forever doomed to inexorably age, with never a chance of rejuvenation? In the case of single stars, the question of attaining a rejuvenated state depends upon its environment; specifically it depends upon its chances of colliding with another star. For most single stars situated within the disc of our galaxy, the chances of a direct collisional interaction with another star are exceptionally small — even over the age span of the Milky Way galaxy (about

12.5 billion years). In galactic and globular clusters, however, the number density of stars can be much higher than that found in the galactic disk, or the solar neighborhood, and this opens up the distinct possibility for close encounters, and even direct star-on-star impacts.

At the present epoch, the closest star to the Sun is Proxima Centauri, to which a distance of 1.3 pc has been deduced from its parallax and proper motion (recall Chapter 1). In terms of the size of the Sun, the present separation between the Sun and Proxima is equivalent to some 29 million Sun diameters. This immediately tells us that in the case of the Sun it is well separated from even its nearest companion — at least, that is, at the present time. What this relative distance number tells us is that the Sun will have to move many tens of millions of times its own diameter before it is likely to encounter another star. If this, or a similar result, holds for a typical star within the Milky Way galaxy, then the likelihood of any two stars colliding since our galaxy formed about 12 billion years ago is effectively zero. The characteristic encounter time t_{enc} between stars can be estimated in terms of the number of stars per unit volume of space N^* (this is the typical number of stars per cubic parsec of space), and the closest approach distance D. The volume V of space swept out by the encounter cross-section πD^2, centered on a specific target star moving with a characteristic velocity, U^*, relative to its surrounding cohort will be $V = t_{enc}\ U^*\ \pi D^2$. An encounter, within the closest approach distance D, is now said to have taken place when $V N^* = 1$. This condition provides us with a formula for the typical encounter time between stars, with a closest approach distance of D,

$$t_{enc} = \frac{1}{U^* D^2 N^*} \tag{3.7}$$

Figure 3.13 shows the variation of typical encounter times with respect to closest encounter distance, D. In the solar neighborhood of the galaxy the characteristic relative speed of stars with respect to the Sun is $U^* = 30$ km/s, and the characteristic number density of stars is $N^* = 0.1$ stars/pc^3. The latter number indicates that there is approximately one star per 10 cubic parsecs volume of space in the solar neighborhood. Indeed, this number gives the characteristic star separation of stars in the neighborhood of the Sun as $\langle S \rangle = 2\,(3 \times 10 / 4\pi)^{1/3} = 2.5$ pc — this is the radius of a sphere containing 10 cubic parsecs of space, and the factor of 2 accounts for each star being contained within the same characteristic volume of space. On the basis of $\langle S \rangle$ being in the order of 2.5 pc, Proxima Centauri, and its companion *a* Cen AB are closer to the Sun than would normally be expected, but this is taken to be just a quirk of our present epoch.

And then came the incident of the dog
(*The Adventure of the Lion's Mane*)

Figure 3.14 indicates what we typically expect to be the case with respect to star encounters within the solar neighborhood. It does not say that collisions between stars do not happen, but it indicates that

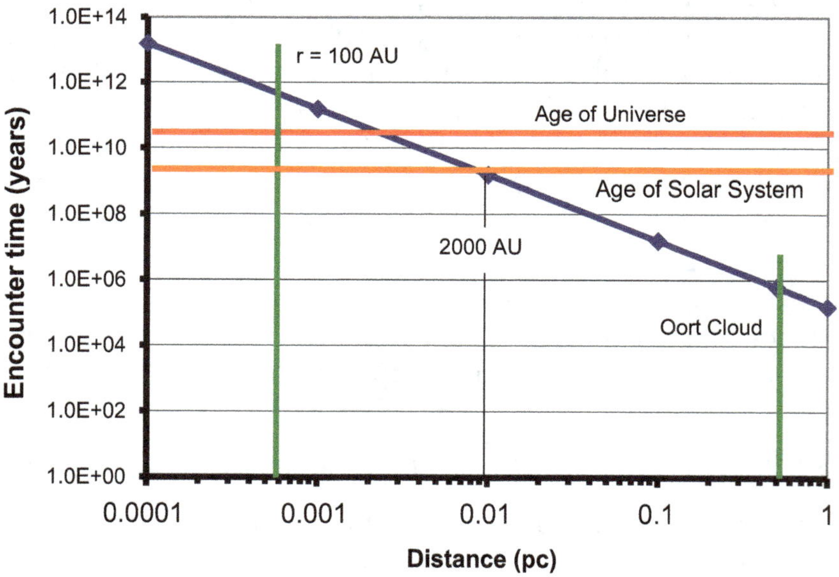

Figure 3.14. Characteristic encounter times for two single stars, as a function of their closest approach distance, in the solar neighborhood. The encounter velocity is assumed to be $U^* = 30$ km/s, and $N^* = 0.1$ stars per cubic parsec. See text for details.

such collisions are, under present conditions in the solar neighborhood, highly unlikely. This very low probability for stellar trysts, at least for the Sun, is important for our very existence. Indeed, a very close encounter between the Sun and even the lowest-mass M spectral type star could be catastrophic with respect to the orbits of the planets. Under typical conditions, Figure 3.14 indicates that over the age of the solar system, some 4.5 billion years, no star will have passed closer than about 2,000 AU of the Sun (that is, no closer than 220,000 Sun diameters). While an encounter at a distance of several thousand AU will place the star well inside of the Solar System (i.e., well inside of the Oort cloud), it will not greatly perturb the orbit of Pluto, nor those of its Kuiper Belt companions situated in the region at about 50 to 100 AU from the Sun. The encounter time interval between encounters as close as 100 AU (or about 13,200 Sun diameters) from the Sun (certainly a Solar System-destroying encounter) is about 0.5 trillion years, but this is a time interval some 36 times older than the current age of the Universe. Indeed, that the Oort cometary cloud still exists about the Sun, after 4.5 billion years, is testament to the fact that no very close encounters have occurred since the Solar System formed — although, for all that, the structure of the Oort cloud has been greatly sculpted by past (reasonably) close stellar encounters. Indeed, it is the case that a star will pass as close as the radius of the Oort cloud (about 100,000 AU) to the Sun once every million years or so, and under these circumstances we do need to be at least thoughtful of the subtle, long-term, forcing of planetary orbital evolution — this is indeed a case of the tail wagging the dog. Depending upon the circumstances, small but commonly occurring forcing terms can drive chaotic variations within orbital characteristics (this, in fact, is exactly how meteorites are delivered to Earth's orbit from the main belt asteroid region). There is no good evidence, however, to indicate that any of the planetary orbits within our Solar System have undergone any significant change because of stellar encounters during the past many billions of years. This may not have been the case during the Solar System's formative years when the Sun, and its newly forming planets, were embedded within a natal cloud where the value of N^* might have been 5 to 10 times higher than in the solar neighborhood today[10]. Even so, we are here, and so, purely upon such grounds, it must be the case that the Sun has never experienced a disruptive, planet scattering encounter with another star. Surveys in which the proper motion and radial velocity data for nearby stars have been collected indicate that the closest (deduced) encounter between the Sun and another star occurred some 70,000 years ago. At this time Scholz's star[11], so-called, passed within 50,600 AU of the Sun. The next very close encounter between the Sun and another star will occur in about 1.3 million years from the present. At this time the K spectral type star Gliese 710 will pass as close as some 10,500 AU to the Sun. At this encounter distance, GL 710 will be well inside the Oort Cloud, and the inner solar system may well be exposed to a distinct uptick in the number of long period comets rounding the Sun.

While close encounters between the Sun and another star is not something that we (as human beings) need worry about, it is the case that very close encounters between stars can occur, and especially so in regions where the stellar number density N^* is high — or at least a lot higher than that in the solar neighborhood. A look at Equation (3.7) indicates that for a given encounter distance, d, the encounter time, t_{enc}, decreases as N^* increases. If N^* is increased by, say, a factor of 100 over that of the present-day solar neighborhood, then we might expect a star to have passed as close as 100 AU to the Sun, on a timescale of some 4.5 billion years (i.e., just once since the Solar System formed). That the star number density in the core of a globular can be a thousand times larger than that of the solar

[10] Not only will a newly forming planet interact with its natal disk, it will also gravitationally interact with any nearby newly forming planets within the natal disc. That planet migration, both inwards and outwards from the host star, is important is evident from both our own Solar System, and the orbital spacings exhibited by multiple-exoplanetary systems.

[11] The trajectory of this star was first described by Ralf-Dieter Scholz (Leibnitz-Institut für Astropysik, Potsdam) in 2015. Another star, HD 7977 is thought to have passed even closer to the Sun than Scholz's star several million years ago, but in this case the encounter distance is only poorly known, with estimates placing it anywhere between 4,000 and 31,000 AU.

neighborhood indicates that very close encounters between stars are highly likely there, and this of course, as we saw in Chapter 2, is where the greatest number of blue stragglers are to be found. Is this a coincidence? No, not at all, and it is the train-wreck formation scenario for blue stragglers that we need to look at next.

3.15. A collision matrix

> *Let us see whether by this purely mental analysis we can get to a finer point*
>
> (*The Adventure of the Three Gables*)

In general, single star on single star collisions are very rare events, but this is not to say that they cannot happen. It is all about location. The more neighbours a specific star has in a given volume of space about it, the more likely a very close encounter, a grazing fly-by, and even a direct hit is likely to take place. The outcome and effects of any such encounters and collisions will depend critically upon the properties of the two stars involved. Table 3.2 provides a rough guide to the possible outcomes when two stars, in different evolutionary phases (and mass), undergo a direct encounter. The table uses a shorthand for various star types and end results — these are:

BD = sub-stellar mass, brown dwarf, with a size comparable to that of planet Jupiter
RD = red dwarf with size a fraction that of the Sun
MS = main sequence star with a size from about 1/10 that of the Sun to some 10 times that of the Sun
RG = red giant star with a size 100 to 1,000 times that of the Sun
WD = white dwarf with a maximum mass of 1.4 $M_\odot$, and size some 1/100 that of the Sun
NS = neutron star with a maximum mass of order 2.5 $M_\odot$, and size some 1/70,000 that of the Sun
BH = black hole (here assumed to be of stellar mass) with an event horizon just of a few kilometers across, ~1/700,000 that of the Sun

Type 1a = Supernova in which a WD undergoes thermonuclear deflagration
KN = Kilonova = nova produced by the merger of two neutron stars
LRN = Luminous red nova = nova-like brightening produced during the merger of two main sequence stars, or the merger of a main sequence star and a red giant
Th-Z = Thorne-Żytkow star = a red giant star powered by the accretion of material onto a centrally located neutron star
Ca-rich = Calcium-rich supernova = supernova showing an unusual spectrum dominated by calcium lines. The origin of this nova type is not entirely clear, but it has been suggested that they are the result of a WD and NS merger

It should be emphasised that Table 3.2 sidesteps many important details, and is rather vague as to the masses of the specific objects that result in a collision or merger. Likewise, it sidesteps issues relating to how the various collisions might proceed. For example, the outcome of a star encounter with a black hole is taken to be the production of a more massive black hole. According to the mass of the black hole and the encounter speed, however, the black hole could simply pass right through the target star, or the target star could be shredded with some of the material being accreted by the black hole and the residual mass being dispersed into space. Multiple encounter outcomes are possible, and the table is intended to be a very rudimentary guide, not a definitive and/or descriptive blueprint.

Table 3.2. Collision outcome matrix for various stellar mass objects. First row and column identify the physical status of the colliding objects. See text for abbreviation details.

	BD	MS	RG	WD	NS	BH
BD	BD/RD	MS/BS	RG	WD/Type 1a	NS/KN	BH
MS	MS/BS	BS	LRN	Type 1a	Type 1a	BH
RG	RG	LRN	LRN	Type 1a	Th-Z	BH
WD	WD/Type 1a	Type 1a	Type 1a	Type 1a	Ca-rich	BH
NS	NS/KN	Type 1a	Th-Z	Ca-rich	KN	BH
BH	BH	BH	BH	BH	BH	BH

From Table 3.2, we see that if two sub-stellar mass brown dwarfs collide and merge, then the outcome could be just a larger mass brown dwarf, or if the total mass of the final post-encounter object is above about 1/10 the mass of the Sun, it could become a bona fide star — indeed, a main sequence red dwarf star. If a brown dwarf is subsumed into a main sequence star, then the end result will be a slightly more massive main sequence star, but now with a mixed interior — that is, it will become a blue straggler. Collisional and merger outcomes involving white dwarfs and neutron stars are likely to result in some form of supernova event, on the basis that these stars have maximum stability masses (recall details in Chapter 2.5 and Table 2.1). Collisions between main sequence stars and red giants will likely produce a luminous red nova (recall Section 3.11). Given a sufficiently slow encounter speed and capture process, an encounter between a red giant and a neutron star might potentially result in the formation of a Thorne–Żytkow star, while a neutron star plus neutron star merger will result in the generation of a kilonova (KN) and an accompanying bright gamma-ray burst. To reiterate, many possible outcomes can result from the collisional encounter between two stars, but the ones of interest here, with respect to the formation of blue stragglers, are those involving main sequence stars. Here, the vital process is one in which a main sequence star gains mass and acquires a well mixed interior, with the latter effect resetting its evolutionary clock. Below, and again in Chapter 4 we discuss several variants on the blue straggler formation theme involving a red giant star and a centrally accreting black hole — such objects being a variant on the Thorne-Żytkow scenario.

3.16. Collision models

A sulphur-match flared blue, then burned with a clear yellow glow
(Adrian Conan Doyle and John Dickson Carr —
The Adventure of the Seven Clocks)

Solving the equations of stellar structure is difficult enough, even with a high-speed computer and copious amounts of memory.

Indeed, the solution methods are mathematically complex and it is especially difficult because many processes work on timescales that are vastly different to each other. Accordingly, astrophysicists make both physical approximations and take mathematical shortcuts to make the solution process robust, but not overly time-consuming. In the situation of modeling two stars colliding, the problem of finding plausible solutions is increased manyfold. Indeed, in the case of collisions, there are process that are taking place on the timescale of minutes to hours, rather than many centuries to many millennia. To circumvent such time differences, astrophysicists have turned to a two-pronged modeling approach [18]. The collision phase is studied with so-called hydrodynamic codes that are specifically designed to look at the evolution of fluid motions, densities, and temperatures. The end results of the collision code calculations are then used as the starting models in a single star stellar evolutionary code. Normally, the starting model for a stellar evolutionary code is that of a chemically homogeneous body descried by a polytropic density profile[12]. In the case of the collision codes, however, the starting model will not necessarily be that of a homogenous protostar, nor will it have a simple polytropic structure. This later condition comes about primarily because of the fact that when two evolved stars collide, the resultant body will have a complicated mean molecular weight profile throughout its interior. Enough said here. The point is that the modeling of a stellar collision is non-trivial, but importantly, not totally intractable.

The hydrodynamic phase is typically modeled according to smooth particle hydrodynamics (SPH) techniques. Under this approximation, a fluid (or gas) is approximated by dividing it up into a set of discrete elements (called particles). The speed, mass, and

---

[12]The full details are not our concern here, but a polytropic structure assumes that the pressure P at each radial location r, within a star is determined by the density ρ, with $P(r) = K \rho^{(n+1)/n}$, where K is a constant and n is the polytropic index. It turns out that zero-age main sequence stars are reasonably well-described by polytropic models with $n = 3$ (where radiative energy transport dominates) and $n = 3/2$ (where convective energy transport dominates).

motion of each particle is then followed numerically with the aid of a kernel function. This latter quantity has a characteristic radius (called the smoothing length), with the idea being that the physical state of each particle is determined according to all the characteristics of the other particles within the domain range of the kernel. In this manner, the kernel is used as a weighting term, with nearby particles having a bigger influence on a given particle's characteristics than particles located further away. There are many computational, time-saving advantages to the SPH approximation, and it has been applied to the study of numerous physical systems in which fluid or gas flows of one kind or another undergo complicated dynamical motion and mixing.

Once the results of the star-on-star collision have been computed by the SPH code, the resultant body is then adopted as the starting model for a standard stellar evolution code, and the long-term effects of any composition mixing and nuclear processing are then followed in detail. By this two-step process, the ability of star-on-star collisions to produce a blue straggler can be examined.

The kinetic energy of two colliding stars is enormous. If we have, say, two Sun-like stars colliding at a relative speed of 5 km/s, the collision energy will be of order 2.5×10^{37} Joules. It would take a Sun-like star some 2,000 years to radiate an equivalent amount of energy into space. Accordingly, with this energy being deposited into the outer layers of the colliding stars, the resultant, merged remnant will be a star that is strongly out of thermal equilibrium. The merged remnant will respond to this disequilibrium by expanding dramatically. Indeed, expanding so much that central fusion reactions are turned off by the concomitant lowering of its core temperature and density. The starting model derived from the SPH collisional models will accordingly be a high-luminosity, low-temperature, bloated stellar mass object — indeed, an object similar to a pre-main sequence star as it settles towards the main sequence and core fusion reactions initiate. The initial collision model, however, will behave differently to that of a pre-main sequence star since it will not be fully convective within its interior, and it will not have a uniform composition. Since fusion reactions are turned off by the collision-induced swelling, the collision product will enter a phase of gravitational contraction on a thermal timescale. Accordingly, the central temperature and density of the collision product will begin to rise again, eventually reaching levels that allow for core fusion reactions to commence (recall Chapter 2.3). The computer models indicate that at this time, the collision remnant essentially settles onto the zero-age main sequence location in the HR diagram, appropriate to that of a star with the combined mass of the two interacting stars.

There are important subtleties, however. The computer models, for example, do not place the collision remnant star on the zero-age main sequence appropriate to that of a fully mixed star. Remember, we are imagining that two somewhat evolved stars collide to produce one new star, and accordingly, the pre-collision stars will have enhanced amounts of helium within their cores (this is because hydrogen fusion reactions were taking place within their central regions prior to the collision). If the collision-produced star had a composition appropriate to that of a fully mixed star, with a composition uniformly averaged of that of the two colliding stars, then its core would be helium-rich. Importantly, computer modeling of stars indicates that the greater the helium mass fraction within the core of a star, the hotter and more luminous its zero-age main sequence location. What all this modeling indicates is that when two stars collide, their interiors become mix in a complicated fashion, and that the resultant, collisional remnant star does not have a uniform (that is, fully-mixed) chemical composition. In this manner, when two stars collide, the resultant collision-produced star is rejuvenated, but it is not fully remade. Also — and this is observationally important — since there is apparently little mixing between the core and envelope compositions of the two colliding stars, the resultant blue straggler is not expected to have a specifically helium-rich envelope, or, indeed, an especially abnormal chemical composition that might be observable in its spectrum. The observational situation with respect to the spectra of blue stragglers is complicated, however, and while many stragglers show no signs of an unusually (compared to companion cluster main sequence stars) chemical make-up, some are, in fact, helium-rich, and others appear lithium- and beryllium- poor.

3.17. Spectral clues

We've laid the family ghost once and forever
(*The Hound of the Baskervilles*)

The most striking example of a sub-population of blue stragglers with a distinctly odd compositional make-up compared to their main sequence counterparts is found in the globular cluster 47 Tucanae (Figure 3.15). Using the 8-m European Southern Observatory Very

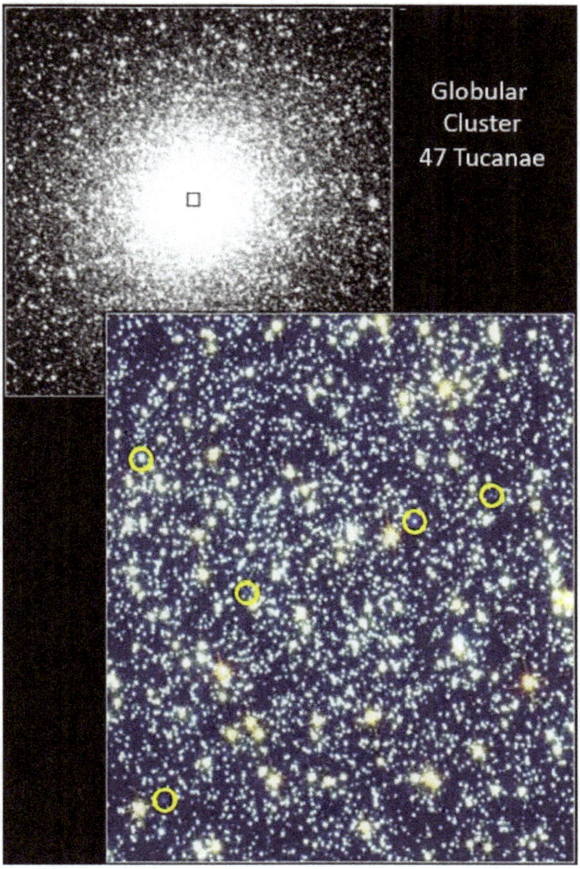

Figure 3.15. The globular cluster 47 Tucanae. The full cluster is seen in the upper image, while the lower image is a zoomed-in view of the cluster core (small central square in the upper image) obtained with the Hubble Space Telescope. Several of the cluster's blue stragglers are indicated by yellow circles. Figure adapted from an image courtesy of NASA.

Large Telescope, F. Ferraro (Università di Bologna, Italy) and co-workers [19] obtained high-resolution spectra of some 43 blue stragglers in 47 Tuc. Of these, 36 had spectral characteristics and abundance measures similar to those obtained for main sequence turn-off mass stars. Of the remaining blue stragglers, however, six showed surface abundances that were significantly reduced in carbon and oxygen (one blue straggler was spinning so rapidly that no abundance analysis was possible). The question of course is, what do these compositional anomalies mean? Recall, star-on-star collisional modeling suggests that there is little mixing between the core and the envelope material, so the surface composition of any collision-produced blue straggler should be similar to that of its cluster neighbours. In the case of mass exchange, however, one would ultimately expect that material from the deep interior of the donor star will be accreted at the surface of the mass-gaining star. Due to the chemical processing, however, of previously running fusion reactions, this late accreted material will potentially appear depleted in elements such as carbon and oxygen. Accordingly, the mass-gaining, resultant blue straggler will have an apparently depleted carbon and oxygen surface layer, and this is what appears to be the case for the six anomalous compositions blue stragglers in 47 Tuc.

Recently, Anne Rathsam (Universidade de São Paulo) and co-workers [20] have highlighted a potential "smoking gun" method for identifying rejuvenated, field blue stragglers. Specifically, they have set about measuring the abundances of lithium (Li) and beryllium (Be) in the outer layers of stars, and comparing these against expectations with respect to population characteristics (recall Chapter 1.2) and galactic motion. The reason for looking at Li and Be is that even at relatively low temperatures, they will undergo rapid destruction by fusion reactions. In the case of lithium (^{7}Li), once the temperature is above a few million Kelvin, proton capture will take place, with P + ^{7}Li ⇒ ^{4}He + ^{4}He + energy. In the case of beryllium (^{7}Be) proton capture results in the formation of boron (^{8}B) which then rapidly decays to ^{8}Be and a positron, with the ^{8}Be further decaying into two ^{4}He nuclei. The point of all this being that if the temperature is only mildly enhanced over those that normally predominate in the outer regions of lower-mass stars, then the abundances levels of lithium and beryllium will rapidly drop towards zero. As a case in point, Rathsam *et al.* discuss their observations of the star HD 65907. This star is an oddity in the sense that its chemical composition is not consistent with expectations, given its motion and location within the thin disc region of our galaxy (recall Figure 1.3). Not only does the star show very low abundance levels of Li and Be, but it also shows excess emission at infrared wavelengths. This later spectral feature is interpreted as indicating the presence of a debris disk surrounding the star. Radial velocity measurements (recall Chapter 1.2), however, reveal no indication of HD 65907 having a binary companion. Furthermore, the isochronal age (recall Figure 2.5) of HD 65907 is estimated to be some 5.6 billion years, and this is significantly lower than the expected age of a typical Population II star within the thin disk of our galaxy — that is, an age of some 11 billion years. Taking all the data together, Rathsam and co-workers suggest that the observed characteristics of HD 65907 are the result of a merger or collision of two old Population II stars [20]. During this merger/collision, not only was the evolutionary clock of the emergent single star reset over that of the colliding/merging components, but the outer regions of the resultant star were heated to such a level that the Li and Be in its envelope were destroyed through the process of efficient and rapid proton capture. Generalizing the results from their study of HD 65907, Rathsam *et al.*, argue that low abundance levels of Li and Be can be used to confirm the generation of blue stragglers as a result of stellar mergers amongst field stars within the galaxy.

A full understanding of the question concerning the appearance of anomalous blue straggler compositions is far from resolved. The numerical models, as they presently exist, do not provide for any clear or distinct predictions separating the various blue straggler generating pathways, and the observations that do exist do not reveal any clear correlation between abundance anomalies, and say rapid rotation, or system age, or formation environment. In this instance, the expected blue straggler composition mystery remains a mystery.

3.18. Post collision evolution

Its not the fall that kills you. Of all people you should know that, its not the fall, its never the fall. It's the landing.

(Spoken by Jim Moriarty: *Sherlock* — TV series: *The Abominable Bride*)

Once fusion reactions begin within the core of a collision-produced blue straggler, numerical calculations indicate that it evolves away from its (reborn) zero-age main sequence location and follows an evolutionary track in the HR diagram that is virtually the same as that of a 'normal' star (formed through the contraction of an interstellar cloud) of the same mass. The differences that are present between the evolutionary tracks of the 'normal' and the collision-produced star are entirely due to the specific compositional inhomogeneities that exist within the interior of the collision-produced star.

While detailed models of colliding stars have been produced, there is much that has yet to be fully understood about the collision process and the resultant collision product. And, indeed, these include an understanding of potential shock heating of the outer layers of the colliding stars and the amount of material lost from the two stars during the collision, along with details of how material and angular momentum are redistributed in the remnant star. This later redistribution is certainly important, since the collision models suggest that the final collision product should be spinning rapidly. This result is partly problematic. While the collision models do indicate that a blue straggler can be formed, they would suggest that all collision-produced blue stragglers should be spinning rapidly. This, however, is not the case, and while some blue stragglers are rapid rotators, most are not. This problem may be ameliorated, however, if one allows for a mechanism that rapidly de-spins a collision-produced blue straggler before it settles into its rejuvenated main sequence phase. One way in which this spin-down might be achieved is through the generation of a strong surface magnetic field in the collision process. This, in fact, is something that Alison Sills (McMaster University, Canada) and co-workers have looked at in some detail, and their modeling results suggest that the formation of a strong, short-lived magnetic field, combined with the ejection of mass during the collision, will effectively act to slow down the collision produced star's spin rate — but the full picture has yet to be completely articulated [21]. The same magnetohydrodynamic numerical models that suggest the production of a strong magnetic field during a star-on-star collision, hint, however, at the possibility of a gas disk forming around the end collision product — especially when off-centre star collisions occur. Indeed, circumstellar disks have been inferred to exist about some blue stragglers, although such disks are not thought to be exclusively the result of an off-centre collision having taken place. Once again, we find that the blue straggler phenomenon continues to challenge the wisdom and modeling abilities of the astrophysicist.

3.19. Vampire stars at the galactic center

We had thought it was some wild tale of foreign parts
(*The Adventure of the Sussex Vampire*)

While a boxer can become punch-drunk and stagger about the ring, an oft-pounded star can rejuvenate itself into a blue straggler, resetting its evolutionary clock by the absorption of lesser stellar neighbors. There is nothing to say that this star rejuvenating situation, vampire-like, cannot occur on multiple occasions. A star, to a certain extent, given a regular diet of appropriate collisions with smaller stars, can live forever (well nearly). The one environment within the Milky Way galaxy where such vampire stars might reside is that situated at its very core. In this central hub region, high-speed collisions are not only likely, but probably a certainty. Indeed, this central region of our galaxy is a busy and crowded place, and at the very center sits a 4.3-million–solar-mass black hole, designated by radio astronomers as Sagittarius A*. For all its mass, the event horizon of this central black hole is smaller than the orbit of Mercury about our Sun, with the observations indicating a diameter of some 52 million kilometers. The great mass of the central black hole dominates the dynamical environment in the core region, and Sgr. A* is not afraid to throw its gravitational influence around. Indeed, it winds up those hapless stars which chance to fall within its gravitational reach into a frenzy of motion, driving them to speeds of hundreds and even thousands of kilometers per second. It is a fast-moving and dangerous stellar mosh pit at the core of our galaxy. Indeed, any star and/or gas cloud passing near to this location had better be wary. Not only might the gravitational swagger of Sgr. A* happily shred such interlopers to pieces, it might also, just as happily, consume them. That we actually know the mass of the Sgr. A* black hole is, in fact, a direct result of the detection and monitoring of (so-called) S-stars passing by and even orbiting about it (Figure 3.16). Monitoring of the motion of S-stars began in the early 1990s, and probably the best studied star is S2. This particular star has been captured into an elliptical orbit about Sgr. A*. By monitoring its motion, and applying Kepler's 3rd law (recall Chapter 1.7) to its deduced orbital characteristics, the mass of Sgr. A* is revealed.

That the environment of Sgr. A* is crowded, and what this crowding might mean, was recently investigated by Sanaea Rose (Northwestern University, Illinois) and Morgan MacLeod (Center

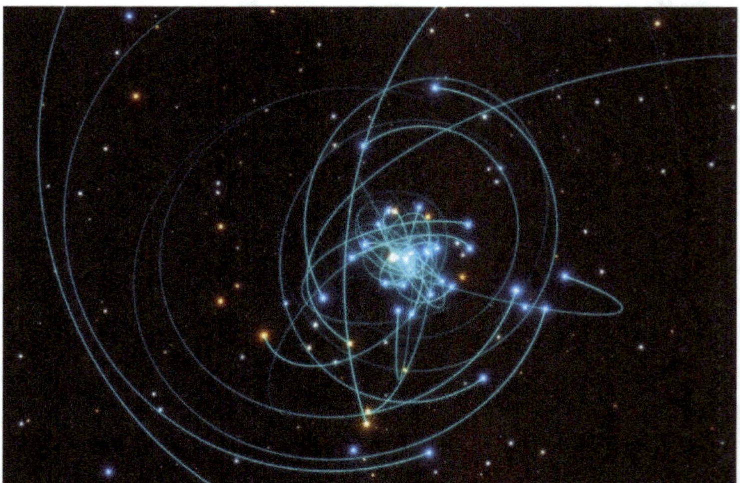

Figure 3.16. Schematic illustration of the paths of S-stars as they move about our galaxy's central black hole (Sgr. A*). Credit: ESO/L. Calçada/Spaceengine.org

for Astrophysics, Cambridge, MA.) [22]. Right at the galactic core, Rose and co-workers argue, the number density of stars will vary with distance from the Sgr. A* according to a relationship of the form $n^*(r) = 1.35 \times 10^6 \, (r \, / \, 0.25)^{-1.5}$. In this manner, in the volume of space contained within a radius of just 1 pc about Sgr. A*, there will be an order of several million stars. In this same volume of space centered on the Sun, the number of stars is two: that is, the Sun and Proxima Centauri. At some 1 AU from the galactic center, the incredible crowding of stars dictates a stellar number density of some $n^* \sim 10^5$ stars per cubic parsec. A glance at Equation (3.7) now indicates that with this value of n^*, and given a characteristic velocity $U \sim 10^2$ to 10^3 km/s, close encounters are going to be very common. With U = 500 km/s, Equation (3.7) indicates encounter times of about 850 yrs, for close approach distances in the order of 1 AU.

While direct collisions are certainly possible, most encounters will be near-misses, with the two stars, as it were, sliding past each other without actual merging. Such encounters, however, are still withering, in that the gravitational interaction between the two stars will result in each star having some of its atmospheric mass stripped away. As with all such lottery interactions, however, there will be winners and there will be losers. Some stars will be gradually, encounter, after encounter, stripped down to a smaller and smaller mass; others will undergo mergers that result in mass gain. It is these latter lottery winners that will resemble blue stragglers, in that the collisions and accumulation of material will result in the enhanced mixing of their interiors and the repeated resetting of their evolutionary clocks. Rose and MacLeod call the mass-gaining, collision-built blue stragglers "zombie stars" since they have become rejuvenated by taking bites out of their neighbors[13].

---

[13] One is tempted to argue that they are more vampirish than zombie-like, since vampires supposedly feed and regain their youth by sucking the blood of others, whereas zombies tend to just rip their food supply apart. It is surely the losers of the grazing encounter that are zombie-like, in the sense that they have lost their mass and thereby their *élan vital* — I may, at this stage, be guilty of overthinking this naming process.

While a supermassive black hole lurks at the core of our Milky Way galaxy, an intermediate-mass black hole appears to lurk at the core of the galactic globular cluster Omega Centauri (recall Chapter 2.8). Are there, accordingly, any hints that the blue stragglers in Omega Centauri are zombie/vampire stars? The answer to this is possibly yes, but the jury is still out concerning a final decision. Recall, the black hole at the center of Omega Centauri was discovered through a proper motion study of stars located at the cluster's core (recall Chapter 2.9). The key stars observed were those with velocities apparently greater than the cluster escape velocity. Rather than escaping directly into intergalactic space, however, these high velocity stars are being held in orbit about a central intermediate-mass black hole. In this sense, the dynamical situation at the center of Omega Centauri mimics that at the galactic center and the S-stars in orbit about Sgr. A*. While the stars used to infer the existence of the black hole within Omega Centauri are identified as 'ordinary' main sequence stars, the cluster is relatively rich in blue stragglers. Interestingly, a study of spin velocities exhibited by blue stragglers within Omega Centauri by Mirko Simunovic and Thomas Puzia (both, at the time of the study, at the Pontificia Universidad Católica de Chile) [23], indicates that the most rapidly spinning blue stragglers are located at the cluster center, while slower spin velocities are measured for those blue stragglers located in the outer regions of the cluster. Indeed, in all globular clusters so far investigated, the most rapidly spinning blue stragglers appear to be centrally located and they typically have high spatial velocities as well (recall Chapter 1.4). This observation is suggestive of the idea that these blue stragglers have formed in a highly dynamic environment, where strong interactions (that is, collisions and mergers) have taken place. In other words, the rapidly spinning blue stragglers located in the core of Omega Centauri are zombie/vampire stars formed and maintained by repeated collisions with other, smaller-mass stars. Well, it's a possibility, but the full story has yet to be written. It is worth pointing out at this stage that there are oddities about Omega Centauri that make it somewhat unusual. Indeed, Omega Centauri is the most massive and the largest of all the globular clusters associated with our galaxy, and it seems relatively clear that it is, in fact, the remnant core of a stripped-down dwarf galaxy. This original structure is probably important, since it is generally thought that black holes are unlikely to form at the centers of globular clusters; it is thought, however, that they might readily form at the centers of galaxies — even dwarf galaxies.

3.20. Binary-Binary encounters

I had come to believe that he was an orphan with no relatives living
(*The Adventure of the Greek Interpreter*)

There is a hybrid collision-merger scenario that can generate blue stragglers. Here the idea is that two binary star systems interact. This is actually a more favorable mechanism for producing a collisional blue straggler than that involving two single stars, since the collision cross-section of a binary is much larger than that of a single star. Figure 3.17 shows an example of the gravitational 'dance' between the components of two interacting binary systems [24]. The two binary (1 and 2) systems have their closest approach at the (0, 0) location in the diagram, and after many close shaves and interactions, the final outcome is a new, tightly-packed binary and two single stars (moving off as shown by the blue arrows). Any number of end results are possible from binary–binary interactions, with the emerging systems being composed of 1) two binary systems, 2) a binary system plus two single stars, 3) a triple star system and one single star, or 4) four single stars. Likewise, a blue straggler can form as a result of the direct collision between two, or even three, of the stars, or through the eventual coalescence of two stars placed into a very close, post-encounter, hard binary. In principle, binary–binary interactions could result in the formation of a blue straggler with a mass three or four times greater that of the cluster turn-off mass, but no such massive blue stragglers have ever been found.

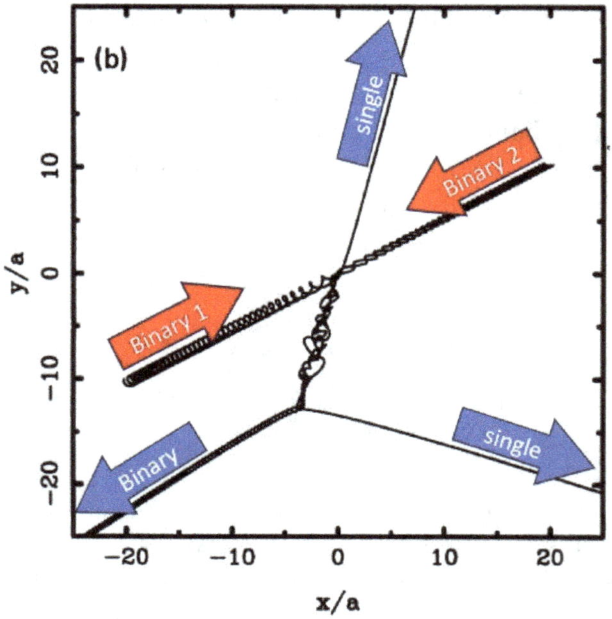

Figure 3.17. Gravitational interaction between two binary systems. The two binary systems start their encounter from the upper right and lower left (red arrows). As they approach each other they undergo an increasingly complex gravitational dance, resulting in the formation of a compact binary (which moves off to the lower left — blue arrow), and two single stars that move off to the top and lower righthand corner in the diagram. Image adapted from Leonard [24].

3.21. Three times the charm

Did anyone know that these proofs would be there?

(*The Three Students*)

Triple star systems (N = 3, recall Chapter 1.3) are not especially uncommon. Indeed, the three closest stars to the Sun form such a three-star system[14]. Likewise, the well-known North Star, or Polaris,

[14]This is composed of the binary system α-Centauri A and B, which orbit each other every 80 years or so, and the single star Proxima Centauri, which orbits α-Cen AB with a period of order 550,000 years.

is a triple system. It is estimated that some 13% of Sun-like (that is spectral types F and G) stars, and some 25% of massive B and O spectral type stars, reside in triple systems. Such systems have intrigued mathematicians, as well as astronomers, for centuries[15], since in order to understand the long-term system dynamics, recourse has to be made to the notoriously difficult three-body problem. The behaviour of such systems admits to no simple description, or easy interpretation, but it does appear that one potential outcome of such system interactions is an increase in the separation of the binary pair and the single star, with a concomitant hardening (that is, reduction in separation) between the two stars in the binary. This interaction is governed according to the so-called Kozai–Lidov mechanism in which the distant (that is, third, single star) companion induces long term oscillations in the eccentricity and inclination of the inner binary's orbit — combining this effect with tidal friction, the two binary stars can then be driven in to a merger. This has relevance to our story, of course, in the sense that the closer the two stars within a binary system approach each other, the more likely it is that they will enter into a mass transference phase and produce a blue straggler.

That triple star systems often evolve to produce hard-binary systems (that is, close binaries which will sooner, or later, experience mass transfer) with a distant, single star outlier has recently been demonstrated by Jonathan Dodd (Leeds University) and co-workers [25] through an analysis of the binary statistics associated with Be stars. B spectral type stars (recall Chapter 1.2, and see Tables 1.1 and 1.2) have masses in the range from about 5 to 15 times that of our Sun, luminosities 1,000 to 10,000 times that of our Sun, and surfaces temperatures 3 to 6 times hotter than that of our Sun. From Table 1.3 we also find that per cubic parsec they are 60 times less common than Sun-like stars, and for every one B spectral type star in the solar neighborhood there are something like 500 M spectral type stars. On the main sequence, the B stars show relatively weak

[15] Issac Newton famously said of the three-body problem, when he applied it to the Earth–Moon–Sun system, that it made his head ache.

Balmer lines along with absorption lines due to ionized helium atoms. A subgroup, of about 20%, of the B stars, however, show something altogether different. They show Balmer emission lines. This spectral characteristic was noted as being 'different' as far back as 1866 when Angelo Secchi first described them. Recall, emission lines are rarely seen in stellar spectra, and the fact that they are seen indicates that something out of the ordinary is happening.

Indeed, Otto Struve suggested, nearly a century ago now, that the Be stars were rapid rotators, and that the observed emission lines originated in a disk of gas orbiting about the star. The disk itself being formed from a stream of gas spun off from the equatorial regions of the B spectral type star. Star rotation was discussed earlier, and Figure 3.3 indicates that for B spectral type stars, the ratio of the observed rotation velocity to the critical rotation velocity is of order $V_{rot} / V_{crit} < 0.3$. For Be stars, however, it was found that $0.5 < V_{rot} / V_{crit} < 0.9$. In other words, many Be stars are spinning close to their upper speed limit. It is this effect that reduces, recall, the effective gravity acting at the star's surface, and this results in mass loss and, in the case of Be stars, the formation of a surrounding equatorial disk. While the rapid rotation argument offers one avenue by which the Be phenomenon can come about, it is thought that other avenues are possible. Just like the blue stragglers, there is more than one way to make a Be star. While rapid rotation is thought to be the key mechanism for Be stars, the origin of the high spin is the problematic point. It could be of primordial origin, resulting from the conservation of angular momentum and the chance rapid rotation of the star's formative molecular cloud. It could be the result of envelope spin-up as a result of momentum transport from a contracting central core, or, indeed, it could result from mass and angular momentum transfer from a close binary companion. It is this close companion hypothesis for Be star origins that Dodd and co-workers looked at with respect to the proper motion (recall Chapter 1.4 and see Figure 1.4) survey data obtained with the *Gaia* spacecraft.

Specifically, Dodd *et al.*, looked at the binary fraction of B and Be stars at different separation levels. Between 0.04 arcseconds and 10 arcseconds separation, it turns out that the binary fraction for B and Be stars are about the same (that is, in the order of 30 percent). At smaller separations, below 0.04 arcseconds, however, the binary fraction of Be stars was found to be significantly smaller than that for B stars (17% versus 28% respectively). The question now is, is this a real effect, and if so, what is it telling us? Well, if we take the effect as being real, the observations indicate that Be stars appear to have fewer close-in companions. What this lower binarity fraction is telling us about Be stars is actually quite subtle, and explained partly upon employing the data reduction technique. To determine binary companionship, Dodd and co-workers measured both long- and short-term proper motion. The idea here is that the long-term proper motion indicates the motion of the center of mass of a binary system (recall Chapter 1.7), while short-term proper motion follows the location of the imaged photocenter — that is the apparent central location of a point source composed of the light produced by two stars in an unresolved binary. By saying unresolved, it is simply meant to say that the two stars are so close together that they look like one star.

The small binary fraction of Be stars at separations below 0.04 arcseconds, can now be interpreted. What appears to be the case, Dood *et al.* argue [25], is that Be stars probably do have close-in binary companions at the same level as ordinary B stars, but in the case of the Be stars they are stripped-down companions too small to be detected through short-term variations in the image photocenter. The rapid spin characteristic of the Be phenomenon is now explained, and rests upon the transfer of mass and angular momentum from a close-in binary companion. As described with respect to the Algol stars, the gainer acquires mass and is spun up through the transfer of material from the donor. The end result is a Be star, with a surrounding disk of gas, spinning at close to its critical rate with a low mass, stripped-down, close-in binary companion. Dodd *et al.* in fact liken Be stars to vampires, since they have fed upon the mass of their hapless companions (recall footnote 13 above).

While the transfer of mass and angular momentum from one star to another can potentially occur in any close-binary system, an additional intriguing fact found by Dodd and co-workers in their proper motion study was that Be stars tended to reside in systems having an additional, wide-separation companion, being part, that is, of a triple or even higher order ($N > 3$) system. The key dynamic point now being that the orbital evolution and 'hardening' of the binary system, which ultimately contains the Be star, is driven by gravitational interactions with the third star in the triple system. Such interactions tend to drive the stars in the binary system closer together and the third, single star to a wider orbital separation with respect to the binary pair.

There is some considerable overlap in the possible pathways by which both Be and blue straggler stars are produced. The former, essentially, being much more massive stars than the latter. Nonetheless, it is possible that at least some of the field blue stragglers, those blue stragglers not situated within a star cluster, are derived from their initial birth within a triple star system. Several examples of such blue straggler formation have been described by Emily Leiner (Illinois Institute of Technology in Chicago) and co-workers [26]. In this case, the stars of interest are found in the old, open cluster Messier 67. While this cluster contains a good number of blue stragglers, two stand out in particular as a consequence of their rapid spin rates. Using observations from the Hubble Space Telescope, the WIYN Telescope at Kitt Peak in Arizona, and the KEPLER Space Telescope, Leiner *et al.* found that not only do these two blue stragglers spin rapidly, they also have white dwarf companions. One of these blue stragglers[16], so-called WOCS[17] 14020, has the appearance of an ordinary Sun-like star but with a rotation period of 4.4 days, and is in orbit about a 0.72 solar mass white dwarf with an orbital period of about 358.9 days. Not only is the very short spin period of WOCS 14020 surprising — the Sun, after all, spins once every 30 days — but so too is the high mass of its white dwarf companion. Acting upon these anomalous values, Leiner and co-workers have set-out to model

[16] Leiner and co-workers actually label these two stars as being blue lurkers since they have estimated masses below that of the 1.3 $M_{\odot}$ turn-off mass of M 67. For all that, they are still blue stragglers. See section 3.23 below.
[17] WOCS stands for the WIYN Open Cluster Survey.

the system's history beginning with it starting out as a triple star system. In this manner, they argue that around some 500 million years ago, the two stars of a close binary situated within a triple star system merged to create a single, relatively large mass (blue straggler) star. This star, in turn, evolved and began transferring material to the (initially) third star in the system. As this mass transfer proceeded, the accreting star was spun up, with the donor (merger-produced) star eventually evolving into a relatively high mass white dwarf. What was three, accordingly became two, and one of the final two is now a rapidly spinning blue straggler.

3.22. Production channel summary

> *Eureka! He cried, his teeth shining through his beard... The problem is solved*
>
> (Arthur Conan Doyle — *The Lost World*)

Some form of summary is now required. Indeed, we have in previous sections identified three essential channels by which blue stragglers might be produced. The mystery of the blue straggler phenomenon has effectively been thrice solved. Figure 3.18 offers a schematic summary of the blue straggler production channels that have been identified. Astronomers have some considerable confidence that the mechanisms identified are correct in principle, and while each scenario has an associated set of unknowns and problem points, theses are considered minor with respect to the bigger picture.

The first channel identified in Figure 3.18 is that involving the collision between two stars. The target star may be single or, indeed, located within a multiple star system, but the point is that as a result of a direct head-on, or even a grazing collision, it acquires additional matter from a donor star. The donor star may be either fully subsumed by the target star in the encounter or considerably reduced in mass; it might additionally leave the collision location and journey onwards into interstellar space.

The second channel seen in Figure 3.18 concerns mass transfer within a close binary system. This is the scenario essentially outlined

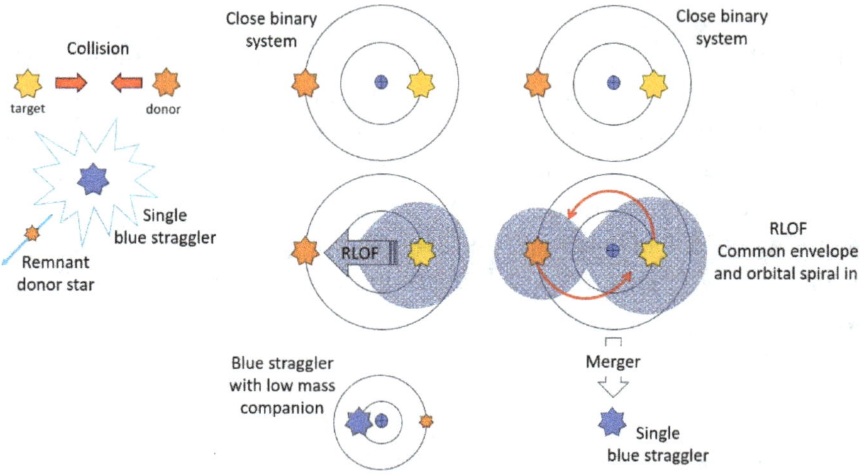

Figure 3.18. Schematic summary of the chief evolutionary pathways for the generation of a blue straggler. On the left, the collision channel is shown, where two stars collide to produce a blue straggler with the ejection of material into the surrounding interstellar space and the possible exit of a much reduced in mass donor star. The target, as well as the donor star, can be single or part of a binary. The middle channel indicates the production of a blue straggler via the process of Roche Lobe Overflow (RLOF) in an evolved close binary system. The resultant binary eventually contains a blue straggler and a low-mass (e.g., a white dwarf) companion. Note, the companion star was originally the more massive component in the binary. In this scenario, the blue straggler may have a high rotation rate and sport an associated circumstellar disk. The third channel (on the right) illustrates the production of a single blue straggler through the merger of two stars within a common envelope binary system.

by W. H. McCrea, and entails the transfer of mass from the initially more massive component in the binary to its initially lower-mass companion. This process proceeds through Roche Lobe Overflow (RLOF), with material traveling from the donor star to the target star through their common L_1 Lagrange point. In this scenario, the final outcome consists of a binary system composed of a blue straggler and a low-mass companion — indeed, the companion will often be a white dwarf.

The third channel illustrated in Figure 3.18 is again associated with the evolution of a close binary system. The binary may be part of a triple system, in which case the presence of a third (single) star may drive the initial orbital evolution of the system. Once again mass exchange via RLOF is critical, but in this case both stars begin to fill their Roche Lobes at about the same time, producing in the process a common envelope system (a W Ursae Majoris variable). Drag forces between the two embedded stars and the envelope then causes the systems orbital separation to slowly decrease, and eventually the two stars merge into one. In this fashion a single blue straggler is produced.

3.23. Channel diversity

> *Thinking, thinking, thinking night and day, what the true meaning of it can be*
>
> (*The Adventure of the Bruce-Partington Plans*)

Summarizing his thoughts on the talks presented at a workshop concerning the observations and theory of blue stragglers held at the Space Telescope Science Institute, Baltimore, in 1992, Mario Livio was to title his proceedings contribution, "blue stragglers: the failure of Occam's Razor?" [27]. The question mark in Livio's title is important, and draws upon a fundamental point about how science proceeds. Occam's Razor is not a scientific principle, however; it is a philosophical device[18]. The basic idea behind Occam's Razor is to not overly complicate explanations, and that the explanation with the least number of assumptions (or impositions) is the best explanation. The term 'best' here is entirely non-scientific, however, and if the history of science tells us anything at all, it is that the simplest explanation for some phenomenon almost invariably needs considerable alteration, adaptation, and complexification as more data is obtained and a better understanding of the physical phenomenon comes about. In his article, Livio identified some 11 suggestions and scenarios for the formation and origin of blue stragglers. At first we

[18] *Entia non sunt multiplicanda praeter necessitatem* [what can be done with fewer is done in vain with more].

might think that this is a bewilderingly large number of ways of making blue stragglers, but then, why should we think this is so? It is only bewildering if we adopt the unnecessary stance that there must be only one simple mechanism operating. In the case of the formation of blue stragglers, Occam's Razor blunts our vision, and indeed, we must accept that there are many ways of forming them. Indeed, multiple channels for producing blue stragglers have been presented, and each channel has its own specific set of necessary conditions for operating efficiently [28]. For collisional production, the density of stars must be high enough, and the system old enough, for encounters to be a likely outcome. For the binary channel to work efficiently, the cluster has to be well-stocked with close binary systems that will undergo mass transfer and common envelope mergers. If the mixing and late formation channels are to operate, then very specific environmental and formation conditions must also be put in place. In general, the total number of blue stragglers within a cluster, N_{BS}, can be written as

$$N_{BS} = N_{late} + N_{mixed} + N_{collide} + N_{merge} + N_{transfer} + N_{other} \tag{3.8}$$

The first two terms in the expression for N_{BS} relate to the number of blue stragglers formed as a result of late or delayed star formation, and homogenous single-star evolution. Our expectation is that N_{late} and N_{mixed} will be small, if not zero, most of the time. The last term N_{other} allows for presently unrecognized channels for blue straggler formation. This term is again likely to be zero in the majority of cases (but see the next chapter). The range and variation in the values for $N_{collide}$, N_{merge}, and $N_{transfer}$ within a given cluster are presently not fully understood to the level that any clear predictions can be made, and indeed, they are time-dependent (that is, dependent on cluster age). The general expectation is that the number of blue stragglers, within a given cluster and at a given specific epoch, will depend almost exclusively on the percentage of close binary systems within the cluster and the evolutionary status of the binaries. This will be further modified by the conditions that determine the level of collisions between stars and binary star systems at the times of cluster core collapse.

Vikrant Jadhav (University of Bonn, Germany) and Annapurni Subramaniam (Indian Institute of Astrophysics) recently looked at the statistics concerning blue stragglers, as identified within open clusters by the *Gaia* astrometry satellite [29]. Among the 670 clusters older than 300 million years for which they had data, 228 of them provide information on 868 blue stragglers. Their first conclusion from the *Gaia* data is that a cluster has to be older than a billion years and more massive than 1,000 $M_{\odot}$ before it can have any blue stragglers. This is consistent with the results given in Chapter 2, Figure 2.7, and Figure 2.9. To tease out the statistics concerning possible formation channels, Jadhav and Subramaniam introduce a quantity called the mass excess, $M_{ex} = (M_{BS} - M_{TO}) / M_{TO}$, where M_{TO} is the cluster turn-off mass and M_{BS} is the estimated mass of the blue straggler. They then argue that if $M_{ex} < 0.5$, the blue straggler most likely formed through low efficiency mass transfer; if $0.5 < M_{ex} < 1.0$ then it most probably formed by a merger (either a collisional event or coalescence within a close binary system); and finally, if $M_{ex} > 1.0$ (when the blue straggler has a mass in excess of twice the turn-off mass) then it presumably formed through multiple mergers and/or collision events. In this manner Jadhav and Subramaniam found that some 54% of blue stragglers were most likely formed via the mass transfer channel, 30% were formed through the collisions and mergers channels, with 16% apparently being the result of multiple mergers. The conclusions that Jadhav and Subramaniam drew from the collected data was that mass transfer within a close binary system is the most common pathway by which blue stragglers are produced. Further to this, they also found that the number of blue stragglers within an open cluster is independent of the cluster's density, radius, and galactocentric location. Likewise, the total number of blue stragglers seen within an open cluster appears to be independent of the cluster's age (once, that is, it is more than about a billion years old — as per our Figure 2.7). For all of that, the central concentration of blue stragglers appears to be directly related to the cluster relaxation timescale (see below).

In principle, the evolution of stars within a cluster can be modeled in a computational simulation. This modeling, however, will need to take into account not only the N-body gravitational interactions of the stars (where N is the number of stars within the cluster), allowing for close encounters, perturbations, and collisions, but it must also follow the evolution of each star within a binary system to account for mass transfer and potential merger events. All of these computational issues are potentially solvable, but to date no fully comprehensive model capable of predicting the blue straggler population for some specified cluster has been fully developed. This latter statement is not to say that attempts at modeling such population numbers haven't been made — they have, but they are generally based upon analytic approximations and guesses involving multiple parametrized terms.

As indicated at the end of Chapter 2, star clusters, whether galactic or globular, are far from static environments –there is a lot going on dynamically. The motions and interactions of the stars in a cluster are described by a series of characteristic timescales. For a given star moving with velocity V in a cluster of radius R, the characteristic crossing time will be of order $t_{cross} = R / V$. The typical time between close encounters between stars is $t_{enc} \sim (V n^*)^{-1}$. It will vary according to the inverse of the speed and the stellar number density as indicated in Equation (3.7). The number of encounters N_{enc} that will typically transpire within a cluster of age $t_{cluster}$, will accordingly be of order $N_{enc} = t_{cluster} / t_{enc}$ — the older the cluster, the more potential encounters a given star will have experienced. The relaxation time, t_{relax}, is the time required for a star to lose any trace of its original orbit, and this will vary according to the number of stars N within the cluster. We need not follow the details here, but a thorough analysis of the relaxation timescale indicates that it varies as $t_{relax} \sim t_{cross} [N / (6 \ln(N/2))]$. Some characteristic values for cluster properties and timescales are shown in Table 3.3.

Table 3.3. Characteristic cluster properties in terms of number (N), size (R), star velocity (V), crossing and relaxation times.

	N	R(pc)	V(km/s)	t_{cross} (yr)	t_{relax} (yr)
Open cluster	100	2	0.5	4×10^6	8×10^6
Globular cluster	10^5	4	10	4×10^5	4×10^8

In general, for the timescales so far discussed, open clusters satisfy the conditions that $t_{cross} \sim t_{relax}$, while for globular clusters, $t_{cross} << t_{relax}$. Furthermore, as discussed in Chapter 2, stars in the core of a cluster tend to have higher speeds than stars in the cluster's outer regions. By interacting with stars outside of the core, the higher velocity stars tend to lose energy and begin to sink ever closer towards the cluster center; this process of core collapse continues until they can find an injection of energy by interacting with the centrally located binary systems — this produces so-called 'hard' or close binary systems at the cluster core and keeps the single stars orbitally energized. The exchange of energy between stars during close encounters not only affects cluster structure and regional number density, it also drives evaporation. That is, once, or if, a star becomes unbound, with the sum of its kinetic and potential energy being greater than zero, the star will escape from the cluster into interstellar space. In this manner, a cluster slowly loses stars into its surrounding environment, and characteristically this evaporation time is taken be of order $t_{evap} \sim 100\ t_{relax}$. Accordingly, with the aid of the values presented in Table 3.3, we recover the result that typically, $t_{evap} \sim 8 \times 10^8$ years for open clusters, and $t_{evap} \sim 4 \times 10^{10}$ years. In this manner, we find that open clusters will tend to evaporate relatively quickly on a galactic lifetime timescale (the galaxy being some 13.5 billion years old), while globular clusters have a dispersal time through evaporation that is some 3 times longer than the present age of the Universe.

In addition to undergoing internal dynamical evolution, a cluster will also undergo external interactions leading to tidal shocks and tidal stripping. Such events will take place when a globular cluster, for example, passes through the mid-plane of the galaxy or the central galactic bulge. For open clusters, tidal shocks will result as a consequence of encounters with giant molecular clouds during spiral arm passages. These external encounter effects will tend to speed up the star evaporation rate.

With all of the above in place, it is clear that cluster evolution is far from straightforward and/or simply to model, and that accordingly, the number of blue stragglers that any specific open cluster of age t might contain, $N_{BS}(t)$, will vary with the outcome of many random variables and events occurring on very different timescales. For all that, Félix Llorente de Andrés, at the Centro de Astrobiología in Madrid, Spain, has developed an analytic model [30] based upon scaling laws, the least squares fitting of data, and observed cluster characteristics to account for the number of blue stragglers that might be seen as a function of cluster age, t. The formula derived for the number of blue stragglers is deceptively straightforward, and indicates that $N_{BS}(t) = f^{3}/(\exp(f / \omega) - 1)$, where f is the ratio of the cluster age to the cluster relaxation time, $f = t / t_{relax}$, and $2 \leq \omega \leq 5$ is a scaling factor partly related to the number of blue straggler formation channels. The formula developed by de Andrés indicates that the expected number of blue stragglers is small when the cluster is very young; indeed, for $t << t_{relax}$, $N_{BS}(t) \sim \omega f^{2} \sim 0$. As the cluster ages, however, $N_{BS}(t)$ will increase up to some maximum value, and then as $t >> t_{relax}$, $N_{BS}(t) \sim f^{3} \exp(-f / \omega) \rightarrow 0$. The maximum number of blue stragglers in a cluster will occur at a cluster age $t = t_{max} \approx 3\omega\, t_{relax}$, at which time, $N_{BS}(t_{max}) \approx (3\omega)^{3} / 19$.

Well, so much for analytic models. However, de Andrés asserts that the formulation for $N_{BS}(t)$ provides a good approximation to the observations for open clusters (indeed, as it should, since it was scaled and constructed according to the observations of such clusters). He also argues that the formulation for the number of blue stragglers can be extended to globular clusters. If this latter assertion is true, then some controversial, but potentially dramatic, results follow.

3.24. Do the numbers add up?

Caution is a scientific attribute, and you must show us your proofs before we can seriously consider your claim

(Arthur Conan Doyle — *The Disintegration Machine*)

Globular clusters, as indicated in Chapters 1 and 2, are composed of some of the oldest known stars. Indeed, the low metallicity (Population II) status of the stars that make up such structures suggests that they are not much younger in age than the Universe itself, having formed and gathered into clusters just a few hundred million years after the formative Big Bang event. The standard cosmological model (including dark matter and dark energy) currently sets the age of the Universe to be some 13.8 billion years, and accordingly, globular clusters cannot have ages that exceed this value. Or can they? The answer to this is controversial, but de Andrés argues that some globular clusters have apparent ages well in excess of 19 billion years. Indeed, using the relationship between blue straggler numbers and cluster age, de Andrés argues that globular clusters such as NGC 5024 (Messier 53) and NGC 5634 could be as old as 21.6 billion years [31]. The question at this stage is how to interpret such numbers. Astronomers have great confidence in the correctness of the standard cosmological model, which gives a canonical age for the Universe of 13.8 billion years, but the standard cosmological model is by no means accepted by all astronomers as being correct. There are numerous alternative cosmological models that have been constructed, and some of these certainly allow for a much older formation epoch than that given by the standard model. The likelihood is that the analytic model (as presently constructed) linking blue straggler numbers to age needs adjusting for such clusters as NGC 5024 and NGC 5634, but the results do, and indeed, should, give us pause for thought.

3.25. Seeing double

There is nothing more deceptive than an obvious fact
(*The Boscombe Valley Mystery*)

Are there hues of blue? Or can a star cluster support multiple generations of blue stragglers? While the observational answer to this question is to be found within a cluster's HR diagram, it is not always clear, from a theoretical perspective, what might be going on. Certainly, some HR diagrams of older clusters appear to indicate multiple epochs of blue straggler formation. Figure 3.19 shows the

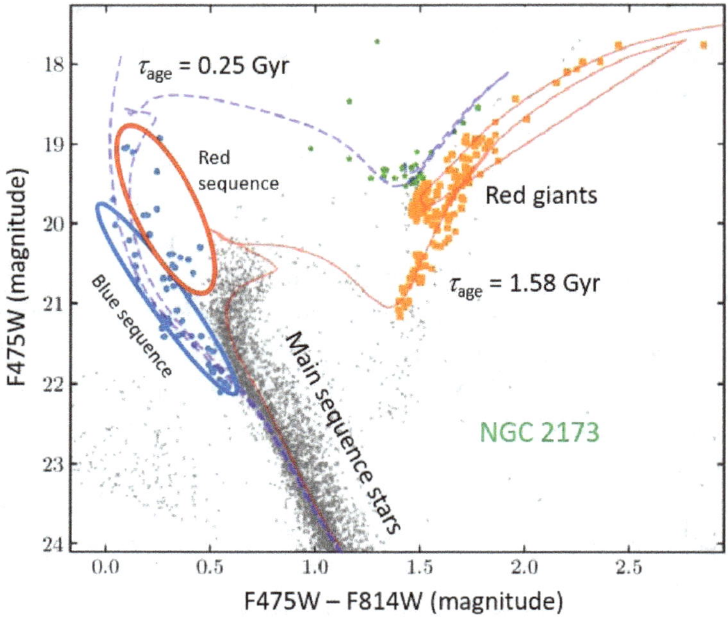

Figure 3.19. Color magnitude diagram for the old open cluster NGC 2173. See text for details and discussion. Image adapted from [32].

HR diagram for the old open cluster NGC 2173, which is located in the Large Magellanic Cloud — a satellite dwarf galaxy to the Milky Way. The analysis behind Figure 3.19 is by Chengyuan Li (Sun Yat-sen University) *et al.*, and the resultant HR diagram is rather busy and complicated [32]. However, the diagram's message can be unpacked according to what we have seen before. First, the diagram is in the form of a color–magnitude diagram, and the individual dots relate to data points gathered with instruments carried aboard the Hubble Space Telescope. Second, the diagram shows a well-defined main sequence and a well populated red giant branch. Fitting an isochrone to the main sequence (red line) indicates a cluster age of just over one and a half billion years. Thirdly, the diagram shows a number of blue stragglers (blue dots) situated beyond the cluster turn-off. Along with the τ_{age} = 1.58 Gyr isochrone, two other isochrones (in blue) for single and equal mass binary star systems with an age of τ_{age} = 250 million years are shown. These younger isochrones provide a good fit to the blue straggler sequence labeled "blue sequence". Finally, however, as noted by Chengyuan Li and co-workers, there are also blue stragglers that lie above the blue isochrones, in the "red sequence" region, suggestive of their being an older (that is, now more evolved) set of blue stragglers with intermittent ages between 0.25 Gyr and 1.58 Gyr.

What are we to make of this apparent blue and red sequence dichotomy? Well, a number of answers are possible, but the full story has yet to be agreed upon by researchers. One suggestion is that the two sequences, the blue and the red, simply relate to two different epochs when the cluster went through a core collapse event — the most recent such event occurring some 0.25 Gyr ago. During core collapse, recall, the density of stars at the core center increases dramatically, and this favors the formation of blue stragglers through the direct collision process. Such an event would be relatively short-lived, and accordingly, produce a tight sequence — the blue sequence — of newly formed blue stragglers all with similar formation age. The red sequence can likewise be associated with an earlier core collapse episode within the cluster. These latter blue stragglers have now developed lower surface temperatures as they evolve towards the red giant region. Alternatively, the red sequence could be the result of blue stragglers being formed via the steady and persistent action of mass-transfer between stars within binary systems. NGC 2173 is not the only cluster to show blue stragglers with apparent blue and red sequences. The color–magnitude diagram for the globular cluster M 30 (recall Figure 1), for example, reveals the same apparent sequencing split of blue stragglers.

That both M 30 and NGC 2173 appear to show blue and red sequences of blue stragglers is suggestive of different origins for blue straggler stars being simultaneously at play. It is also puzzling. The reason why a clean interpretation of the blue straggler sequences, as a consequence of two formation channels being in operation, is somewhat problematic follows from the fact that the two clusters have distinctly different ages: NGC 2173 is some 1.58 billion years old; M 30, in contrast, as a globular cluster is some 12.9 billion years old. Accordingly, while we might well expect M 30 to be old enough for multiple core collapse events to have occurred, NGC 2173, in contrast, as a relatively young cluster, is unlikely to have undergone multiple core collapse episodes (it is just too young and too poorly populated). In other words, as Li and co-workers conclude, virtually all of the blue stragglers in NGC 2173 were likely formed via mass transfer and/or coalescence within binary systems.

A clear explanation for the multiple blue straggler sequences observed within some cluster color–magnitude diagram's remains obscure, and the problem continues to defy a complete understanding. One possibility for future investigation is to see if the dichotomy in the spatial distribution of blue stragglers, as discussed by Mapelli *et al.* (recall Figure 2.11), holds true with respect to blue and red sequence blue stragglers: are the blue sequence blue stragglers preferentially located in the cluster core (where the collision formation mechanism should dominate), and are the red sequence blue stragglers preferentially located within the outer cluster region (where the binary formation channel is likely dominant)? Such studies have yet to be performed, and the fact that such questions have yet to be answered underscores the point that while much of the blue straggler mystery has been resolved, there is still much work to be done. A step in resolving the bifurcation sequence issue, however, has recently been undertaken by Li Wang (Sun Yat-sen University, China) and co-workers [33]. In this new study, a grid of binary stellar evolution models has been constructed, and the authors conclude that the red sequence of blue stragglers in NGC 2173 are associated with ongoing mass transfer under conservative conditions. They predict that the blue sequence, in contrast, is composed of blue stragglers that have completed a phase of non-conservative mass transfer and have associated white dwarf companions. Furthermore, Wang *et al.* conducted a detailed N-body simulation under conditions applicable to NGC 2173, and found that no star-on-star collisions were at all likely given the cluster's relatively young age.

3.26. Stragglers of many colors

The permanency or variation of star-tints requires still closer observation before any correct deductions can be drawn
(W. H. Smyth — *Sidereal Chromatics* — published in 1864)

The existence of blue stragglers first became apparent because of the fact that they are located within star clusters. They stood out by being brighter and hotter than the nominal cluster turn-off location — indicating that they have masses greater than that of the cluster turn-off mass, M_{TO}. As seen in the chapters thus far, they are formed as a result of either collisions between single stars, or mass transfer and/or mergers within binary systems. Importantly, however, whatever channel they might have formed by, the blue stragglers end up as main sequence stars with evolutionary clocks that have been reset. It is entirely possible, however, that the merger and/or collisional processes that resulted in the formation of a blue stragglers could produce a straggler with a mass less than that of the cluster turn-off mass. In this situation, the main sequence clock of a star can be reset by mixing and merging, but the resultant straggler is hidden within the crowd of stars that delineate the main sequence in the cluster's colour-magnitude (HR) diagram. Here we have a blue lurker. Like a tree hidden in a forest, there is nothing to specifically distinguish a blue lurker from any other main sequence star that might reside in the cluster. Ultimately, however, a blue lurker might, through the resetting of its evolutionary clock, become a blue straggler, since sooner or later, its similar-mass (but older in an evolutionary sense) main sequence stars will evolve off the main sequence.

As discussed in Chapter 2, once a star ends its main sequence phase, it swells up to become a red giant prior to the initiation of helium fusion reactions within its core. At this red giant stage, within a cluster HR diagram, the stars are located on the giant branch, and are characterized by being cool, large and luminous. The time it takes a star to transform between its main sequence location in the HR diagram to that of its red giant position is very short, being achieved within a few hundreds to thousands of years. This is a dramatically short time, especially when compared to the main sequence lifetime of the star, which for low mass stars is measured in tens of billions of years. Accordingly, in the HR diagram, there is a region between the main sequence and the red giant branch that appears to contain very few stars. This region is called the Hertzsprung[19] gap (HG), and is readily seen in Figure 1.17. It is not that stars cannot have luminosities, radii and temperatures that place them within the HG, it is just that they have them for such a short period of time that statistically we would not expect to see them in such a position. Effectively, the transition from being an end–main-sequence star to becoming a red giant is so rapid that it appears almost as an instantaneous jump in the star's position in the HR diagram (see Chapter 4). Imagine the surprise of astronomers, therefore, when in the HR diagram of some open clusters they actually observed stars in the region of the cluster's Hertzsprung gap. These stars have become known as yellow stragglers. That is, they appear to be straggling with respect to achieving full red giant status. And, indeed, McCrea predicted that such stars should exist in his 1964 study (recall 'consequence 5' in Section 3.9 above).

Figure 3.20 shows the magnitude–color HR diagram for the open cluster NGC 2447 (also cataloged as Messier 93). This compact cluster is about 3 pc across, about 1,000 pc distant from the Sun in the galactic plane, with a total mass of about 700 $M_{\odot}$, and an estimated age of some 630 million years. To the left in the diagram is the main sequence, and the extended blue straggler region is indicated for those stars above the cluster turn-off. To the upper right is the red giant branch, and the diagonal dashed lines indicate the region corresponding to the Hertzsprung gap. For NGC 2447, some four yellow stragglers have been identified (so far, along with at least five blue stragglers) [34].

The evolutionary status of yellow stragglers remains unclear. M. da Silveira (Observatório Nacional, Brazil) and co-workers [34] argue, on the basis of a detailed spectroscopic study, that all the

[19] This is the same Ejner Hertzsprung who first constructed the HR diagram — see Chapter 1.

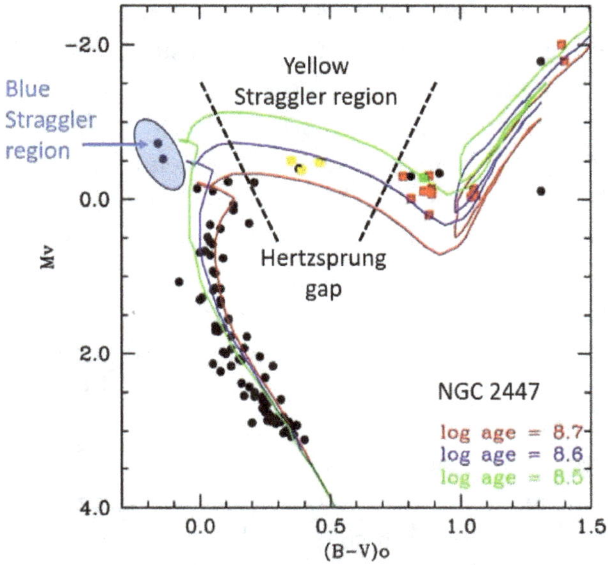

Figure 3.20. HR diagram for the galactic, open cluster NGC 2447. The blue straggler region is indicated towards the upper left in the diagram, with the region containing the yellow stragglers falling between the boundaries (dashed lines) of the Hertzsprung gap. Also shown are a set of three isochrones for cluster ages between 316 million and 500 million years. Diagram adapted from da Silveira *et al* [34].

yellow stragglers in NGC 2447 are binary systems composed of a red giant and an A spectral type main sequence star. In contrast, other researchers suggest, along the lines predicted by McCrea, that they are evolved blue stragglers, slowly making their way to the red giant branch. Yet, others have also suggested that they are the result of mass transfer within a binary system, with the straggler gaining mass from an evolved asymptotic giant branch star — a star that will eventually become a low-mass white dwarf. Other observations indicate that at least some yellow stragglers show evidence for being magnetically active. Perhaps, therefore, the yellow stragglers are actually sub-giant stars supporting a strong magnetic field. In this manner, the magnetic field suppresses the convective efficiency in the outer layers and produces a large covering of star spots — these two effects effectively reducing the temperature and luminosity of the host star.

Currently there is no clear consensus as to what the yellow stragglers are telling us, and this lack of consensus even extends to the fact that some researchers refer to them as red stragglers. The latter label is applied according to the notion that it is the attainment of red giant status that is being delayed. This being said, the label 'yellow straggler', which reflects their intermediate color (that is, temperature) between the blue of the main sequence and the red of the giant branch, is perhaps a more appropriate label for those stars found dallying in the Hertzsprung gap. This is potentially further helpful in light of the recent theoretical prediction that a distinct class of objects, more appropriately labelled as red stragglers, might exist [35]. These red stragglers will be discussed in more detail in Chapter 4, but the basic idea behind such objects is that they are accretion-powered giant stars. In this sense, they are taken to be stars powered by accretion onto a centrally located low-mass black hole. These stars linger on, or near to, the red giant branch since their luminosity is provided for through accretion, and, because of this, the onset conditions for central helium fusion reactions are supressed.

3.27. Out in the fields

Not invisible, but unnoticed, Watson, you did not know where to look

(*A Case of Identity*)

While we have introduced blue, red, and yellow stragglers, and even blue lurkers, there is one more kind of straggler to mention; these might be called invisible stragglers or stragglers of many colors. As such, these stars are blue stragglers seen outside of a cluster environment. More commonly called field blue stragglers (recall Table 2.2), these stars are difficult to identify since they have no neighboring stars, of the same formation age and composition, against which to gauge their straggling nature. Indeed, they simply look like main sequence stars. Recall, however, the study by Rathsam *et al.* [20], where it was argued that low Li and Be abundances are an indicator of stellar mergers. This group of field stars must exist since the channels leading to blue straggler formation can certainly operate outside of a star cluster environment. This being said, however, field blue stragglers are most likely produced through the mass transfer and binary merger channels, rather than the collision channel, since they are typically located within regions of low star density.

By carefully delineating specific regions within a color–color (that is, HR) diagram (recall Figure 1.2), it is possible to pick out field stars that are potential blue stragglers. By combining such photometric survey data with spectrographic information on the same stars, a winnowing down of candidates can be done, and the spatial distribution of field blue stragglers within the galaxy can be investigated (recall Figure 1.3). Just such a process was carried out by Rafael Santucci (Universidade de São Paulo, Brazil) and co-workers [36] using data gathered from the Sloan Digital Sky Survey (SDSS) and the Sloan Extension for Galactic Understanding and Exploration (SEGUE) data set. Indeed, from the survey data, Santucci *et al.* extracted information on the positions and properties of some 8,001 field blue stragglers (FBS) along with that relating to 4,796 blue horizontal branch (BHB) stars. From this data set they set about investigating the galactic variation in the ratio of the number of field blue stragglers compared to blue horizontal branch stars: $F = N_{FBS} / N_{BHB}$ (for reference, $F \sim 0.5$ to 1.0 for globular clusters, and $F \sim 2$ to 3 for open clusters). Given that field blue stragglers are likely all produced through mass transfer or mergers within binary systems, variation in the value of F, with respect to galactic coordinates, will tell us something about the primordial distribution of binary stars.

Accordingly, with reference to Figure 1.3, the maximum value of $F \sim 4.0$ is found within the thick disc of the galaxy at heights between 2 and 5 kpc above and below the galactic plane. In the inner galactic halo, the F value steadily falls to between 1.5 and 2.0. The F value is close to 0.5 in the outer galactic halo. The F value in the galactic plane (that is within the disk), is in the order of 1 at a distance of some 5 kpc from the galactic center. The numbers ratio then increases to a maximum value between 3 and 4 at a distance of some 12 kpc from the galactic center, gradually falling thereafter to an order of 1 at distances beyond about 25 to 30 kpc from the galactic center. These results indicate that field blue stragglers, our proxy for the distribution of primordial close binary systems, are preferentially confined within the thick disk region of the Milky Way galaxy, at radial distances smaller than 12 to 15 kpc of the galactic center. With respect to the thin and thick disc variation, the increase in the relative number of blue stragglers is consistent with the notion that the binary fraction of stars is inversely related to metallicity — the lower the metallicity, the greater the relative number of binary systems. The observed fall-off in the number of field blue stragglers close-in towards the crowded regions of the bulge and galactic center, which are locations that are predominantly composed of old Population II stars (like the thick disk), is interpreted according to the likely disruption of primordial binary systems through repeated gravitational encounters with other stars. Indeed, as seen earlier, this is a region where zombie/vampire (blue straggler) stars are to be found (recall Section 3.19). The dearth of blue stragglers in the outermost halo is taken to be an indication that it simply never contained many binary systems.

Further to this, however, Santucci and co-workers found, upon a closer inspection of the orbital dynamics and compositions associated with those field blue stragglers situated within the outer halo, that perhaps 5% of them are extragalactic in origin. That is, they have been accreted by our galaxy in mergers with other smaller dwarf galaxies in the distant past[20]. Some blue stragglers can indeed trace their origins to "a long time ago, in a galaxy far, far away." The full interpretation of such large-scale survey results requires more analysis, but what Santucci and co-workers have shown is that the relative number of field blue stragglers decreases with both

[20] Santucci and co-workers found, for example, an excess of blue stragglers and horizontal branch stars in the region of what is called the Sagittarius stream. This region is composed of stars that have been tidally stripped and captured from the Sagittarius dwarf spheroidal galaxy (Sgr dSph), which is a satellite galaxy to the Milky Way. The orbit of Sgr dSph takes it above and below the Milky Way's galactic poles at a radial distance of about 20 kpc, and it periodically passes through the galactic plane. Eventually, over the next several billion years or so, the entire dwarf galaxy will be stripped and its stars subsumed by the Milky Way.

distance from the galactic center and height above and below the galactic plane, and this, in turn, informs us about the variation in the galaxy's primordial structure and distribution of binary star systems.

It opens a pleasing field for intelligent speculation
(*The Adventure of the Red Circle*)

References

[1] J. C. Wheeler. Stars with anomalous mass — Is there funny business in the main sequence. *Comments on Modern Physics, Part C — Comments on Astrophysics*, 8, 133–147, 1979.

[2] H. Saio, and J. C. Wheeler. The evolution of mixed long-lived stars. *The Astrophysical Journal*, 242, 1176–1182, 1980.

[3] M. Beech. *Introducing the Stars: formation, structure and evolution.* Springer Nature, Switzerland AG. 2019.

[4] A. Billi, *et al.* Rotational velocities of blue straggler stars in the globular cluster M 55. *Astronomy and Astrophysics*, 690, A156, 2024. https://doi.org/10.1051/0004-6361/202450593.

[5] A. Mucciarelli, *et al.* Spinning like a blue straggler: the population of fast rotating blue stragglers in ω Centauri. *The Astrophysical Journal*, 797, A43, 2014.

[6] F. Ferraro, *et al.* Fast rotating blue stragglers prefer loose clusters. *Nature Communications*, 14, A.2584, 2023. https://doi.org/10.1038/s41467-023-38153-w.

[7] H. Boffin. Mass transfer by stellar wind. In: H. Boffin, G. Carraro, G. Beccari (Eds) *Ecology of Blue Straggler Stars*. Astrophysics and Space Science Library, 413. Springer, Berlin, Heidelberg. https://doi.org/10.1007/978-3-662-44434-4_7.

[8] P. Eggleton. Approximations of the radii of Roche Lobes. *The Astrophysical Journal*, **268**, 368–369, 1983.

[9] W. van Rensbergen., and J-de Greve. On the Modeling of Algol-Type binaries. *Galaxies*, 9, A9, 2021.

[10] N. Gosnell, *et al.* Constraining Mass Transfer Histories of Blue Straggler Stars with COS Spectroscopy of White Dwarf Companions. *The Astrophysical Journal*, 885, A45, 2019. https://doi.org/10.3847/1538-4357/ab4273.

[11] W. H. McCrea. Extended main-sequence of some stellar clusters. *Monthly Notices of the Royal Astronomical Society*, **128**, 147–155, 1964.

[12] N. Soaker, and Romuald Tylenda. Modelling V838 Monocerotis as a Mergeburst Object. In, *The Nature of V838 Mon and its Light Echo*. R. Corradi and U. Munar (Eds.). ASP Conference Series, 30, 280–290, 2006.

[13] A. Retter, *et al.* The planets capture model of V838 Monocerotis: conclusions for the penetration depth of the planet(s). *Monthly Notices of the Royal Astronomical Society*, 370, 1573–1580, 2006. https://doi.org/10.1111/j.1365-2966.2006.10585.x.

[14] R. Tylenda, *et al.* V1309 Scorpii: merger of a contact binary. *Astronomy and Astrophysics*, 528, A114, 2011. https://doi.org/10.1051/0004-6361/201016221.

[15] A. Frost, *et al.* A magnetic massive star has experienced a stellar merger. *Science*, 384, 214–217, 2024.

[16] B. Lee, *et al.* Search for exoplanet around northern circumpolar stars: Four planets around HD 11755, HD 12648, HD 24064, and 8 Ursae Minoris. *Astronomy & Astrophysics*. **584**: A79, 2015. https://doi.org/10.1051/0004-6361/201527076.

[17] H. Chen, *et al.* A high-velocity star recently ejected by an intermediate-mass black hole in M 15. *National Science Review*, 12, nwae347, 2025. https://doi.org/10.1093/nsr/nwae347.

[18] J. Lombardi, *et al.* Stellar collisions and the interiors of blue stragglers. *The Astrophysical Journal*, 568, 939–953, 2002.

[19] F. Ferraro, *et al.* Discovery of carbon/oxygen-depleted blue stragglers in 47 Tucanae: the chemical signature of a mass transfer formation process. *The Astrophysical Journal*, 647, L53-L56, 2006.

[20] A. Rathsam, J. Melendez, and A. Karakas. Beryllium: the smoking gun of a rejuvenated star. *Astronomy and Astrophysics*, 683, A26, 2025. https://doi.org/10.1051/0004-6361/202451197.

[21] A. Sills., *et al.* Evolution of stellar collision products in globular clusters: Head-on collisions. *The Astrophysical Journal*, 487, 290–303, 1997.

[22] S. Rose, and M. MacLeod. Collisional shaping of nuclear star cluster density profiles. *The Astrophysical Journal Letters*, 963, L17, 2024.

[23] M. Simunovic, T. Puzia, and A. Sills. The blue straggler star population in NGC 1261: evidence for a post-core-collapse bounce state. *The Astrophysical Journal Letters*, 795, L10, 2014.

[24] P. Leonard. Stellar collisions in globular clusters and the blue straggler problem. *The Astronomical Journal*, 98, 217–226, 1989.

[25] J. Dodd, *et al. Gaia* uncovers difference in B and Be star binarity at small scales: evidence for mass transfer causing the Be phenomenon. *Monthly Notices of the Royal Astronomical Society*, 527, 3076–3083, 2023. https://doi.org/10.1093/mnras/stad3105.

[26] E. Leiner, *et al.*, https://science.nasa.gov/missions/hubble/nasas-hubble-tracks-down-a-blue-lurker-among-stars/.

[27] M. Livio. Blue stragglers: The failure of Occam's razor? In: H. Boffin, G. Carraro, G. Beccari (Eds) *Ecology of Blue Straggler Stars*. Astrophysics and Space Science Library, 413. Springer, Berlin, Heidelberg. https://doi.org/10.1007/978-3-662-44434-4_7.

[28] N. Leigh, A. Sills, and C. Knigge. An analytic model for blue straggler formation in globular clusters. *Monthly Notices of the Royal Astronomical Society*, 416, 1410–1418, 2011.

[29] V. Jadhav., and A. Subramaniam. Blue straggler stars in open clusters using *Gaia*: dependence on cluster parameters and possible formation pathways. *Monthly Notices of the Royal Astronomical Society*, 507, 1699–1709, 2021.

[30] F. de Andrés. A model of the dynamics of open clusters: time-scales, core collapse and blue stragglers. *American Journal of Astronomy and Astrophysics*, 9, 52–66, 2022.

[31] F. de Andrés. Is the universe older than commonly accepted? Observations of JWST and globular clusters have their say. https://arxiv.org/html/2401.11549v1.

[32] L. Chengyuan., Deng, L., R. de Grijs., D. Jiang., and Y. Xin. An Unexpected Detection of Bifurcated Blue Straggler Sequences in the Young Globular Cluster NGC 2173. *The Astrophysical Journal*, **856**, 25, 2018. https://doi.org/10.3847/1538-4357/aaad65.

[33] L. Wang, *et al.* Unveiling bifurcated blue straggler sequences in NGC2173: insights from binary evolution. https://arxiv.org/html/2503.19966v1.

[34] M. da Silveira, C. Pereira, and N. Drake. Red giants and yellow stragglers in the young open cluster NGC 2447. *Monthly Notices of the Royal Astronomical Society*, Volume 476, Issue 4, June 2018, Pages 4907–4931, https://doi.org/10.1093/mnras/sty265.

[35] E. Bellinger, *et al.* Solar evolution models with a central black hole. *The Astrophysical Journal*, 959, 113, 2023. https://doi.org/10.3847/1538-4357/ad04de.

[36] R. Santucci, *et al.* The frequency of field blue-straggler stars in the thick disk and halo system of the galaxy. *The Astrophysical Journal*, 801, A116, 2015. https://doi.org/10.1088/0004-637X/801/2/116.

Chapter 4
And finally ...

From a drop of water, a logician could infer the possibility of an Atlantic or Niagara without having seen or heard of one or the other. So all life is a great chain, the nature of which is known whenever we are shown a single link of it.

(*A Study in Scarlet*)

There is a narrow, razor-thin, boundary that separates the scientifically unknown and exploratory from the sticky realm of pseudo-science and delusion. It is probably fair to say that no one specifically sets out to develop a theory of pseudo-science, but the content of this final chapter is more speculative, more extreme, and seated closer to the edge of the scientifically unknown. Arthur C. Clarke famously advised that to go beyond the known one has to push into the unknown, and this is my aim in this final chapter. Where, indeed, might the idea of blue stragglers take us? Where, in the sense of Sherlockian water-drop logic, might the dreamer's mind go, and what wonders might such mind-pictures reveal, given the existence of blue stragglers?

4.1. The final problem

The problem presents features of interest ... I may even say very exceptional features of interest.

(*The Crooked Man*)

There is a well-known saying that intones the idea of 'go big, or go home'. From our investigations into the blue straggler mystery, however, this saying might be reinterpreted by our far distant descendants to 'go big, and stay home'. At stake in our discussion here (in our singular, Earth-bound situation) is the very future of humanity, with the final problem being recognized as what to do about the ever-increasing luminosity of the Sun as it evolves through its main sequence phase and beyond. The Sun's luminosity is increasing by about 1 percent per billion years [1], and while this does not seem like much, within two to three billion years from the present, the Sun's energy output will be so high that Earth's oceans will begin to evaporate. With no oceans and a blistering Sun in the sky, life on Earth, in all of its wonderous forms, will come to an end. Indeed, life on Earth (the biosphere) will, under this standard scenario, end several billion years before the end of the Sun's main sequence phase. At this stage, what passes for humanity (that is, our very different and very distant descendants) might contemplate the abandonment of Earth, and even the solar system, in search of a new planet upon which to live. Or, of course, they might stay in place and perish with the rest of biosphere[1]. They might also simply abandon terrestrial life altogether and adopt an interstellar, gypsy-like existence. What if, however, the choice is made to stay in place, and yet not accept an inevitable demise? Are there any additional scenarios that can make such a hard choice not only possible but reasonable? If we are prepared to think big and assume an ever-growing skill-set for our (assumed) very, very distant descendants, then theoretically there are ways to avoid the untimely collapse of the biosphere. Indeed, given the remit of this final chapter, we are allowing ourselves to question the very phrase "under the standard scenario" with respect to the Sun's evolution. The question being, is it possible to 'change' or

[1] In another context, I have suggested that technically-advanced civilizations might completely disconnect themselves from nature and adopt an existence deep under the surface of their home planet. Such Schrödinger worlds, as I have called them, will accordingly be both dead and alive — dead on the surface, but (invisibly) thriving and full of life deep below the surface.

'engineer' the Sun so that it does not evolve a higher luminosity and/or move towards a giant configuration as it ages? In some sense, therefore, the question is, can the Sun be turned into some form of blue straggler, that is a rejuvenated star, without significantly increasing its mass or luminosity? Clearly, we (that is, humanity) are currently in no position to alter the Sun in any way whatsoever, but we can run numerical simulations of stars under non-standard conditions (recall, for example, Chapter 3.3) [2]. Here it will be the equations of stellar structure that will act as our guide to what might be engineered in the far distant future by our far distant descendants.

We need not specifically write down the equations of stellar structure here; suffice to say they are a set of interconnected differential equations that determine the run of such variables as temperature, pressure, density, energy generation, composition, and mass as a function of radius — as described earlier in Chapter 2. There is no one or unique solution to these equations; there are indeed an infinite number of possible solutions, and these can only be corralled by adopting various physical assumptions and applying specific boundary conditions. This being said, if one delves into a higher-level book on stellar astrophysics [3], one will find mention of the so-called Vogt—Russell (VR) theorem. This is a somewhat aggrandized label, since it is not an actual theorem, nor does it have any mathematical proof of statement, but what the VR 'theorem' sets-out to say is that a star's structure, as described by the equations of stellar structure, is fully and uniquely determined by specifying its total mass and its internal chemical composition. There is a caveat to the VR 'theorem' in that the statement is supposed to hold true provided that the opacity, energy generation, and pressure, at any point within a star, are taken to be functions of the density, temperature, and chemical composition alone (recall Chapter 3.3.1). This is a lot to take in, but what the VR 'theorem' really says is that to solve the equations of stellar structure, a specific set of boundary conditions must be given from the outset. The mathematical difficulty with this statement, however, is that some of the boundary conditions are set at the center of the star, some at the surface, and some at every point within its interior. Problematically for its label as a theorem, however, is that there is no mathematical proof that a unique solution to the equations must result once the boundary conditions have been specified. This further informs us that there is no analytic solution to the full set of equations, and that a numerical scheme, based upon dividing the stellar model into a series of concentric shells, must be developed, with the equations being solved on each of the shell boundaries. In this manner, a stellar model is really just a printout of numbers representing physical terms (mass, composition, pressure, density, and temperature) at a finite number of radial locations r between the center ($r = 0$) and surface ($r = R$). The more shells that the computer can handle, the more precisely will the internal structure of the star be described. Solutions to the equations of stellar structure can certainly be found numerically, and usually all goes well in the computational process. Even so, more than one solution for the internal structure and configuration of the star might still exist. Indeed, the existence of the Hertzsprung Gap (recall Section 3.24 on yellow stragglers) comes about partly because of the Schoenberg–Chandrasekhar limit[2], where a stellar model effectively has two equilibrium solutions. In this case, for a set stellar mass and composition, one solution places the star at its terminal main sequence location, while the other places it at the base of the red giant branch.

Further to all of the mathematical issues concerning uniqueness and potential multiple solutions to the equations of stellar structure, there are related (in)stability conditions and physical approximations made for the mode of energy transport that must be taken into account. In the latter case, the question is whether the energy is transported outward by radiation, convection, and/or (under extreme compression conditions) conduction. Furthermore, the opacity must

[2]Again, the full details are not our main concern here, but the Schoenberg-Chandrasekhar (SC) condition relates to the maximum pressure that an isothermal core can support once fusion reactions have exhausted the available supply of hydrogen at the end of a star's main sequence lifetime. Once the SC condition is breached, the core must contract and the entire star must find a new equilibrium state by adjusting its radius and temperature — that is, by becoming a red giant.

also be defined at each point within the star, since this determines how strongly the radiation interacts with, and is impeded by, the medium through which it is moving. Each of these issues raises complex physical problems. If the energy transport is by convection, for example, then the efficiency of this process has to be modeled by some *ad hoc* approximation theory. Likewise, the opacity changes as a complex function of composition, temperature, and density, and there is no simple (or even complex) general formula to describe how these changes proceed. There are, of course, ways and means to get around these problems, but they involve the imposition of various physical assumptions, parameterizations, and simplifications. Accordingly, solutions to the equations of stellar structure can be found, and the model parameters can be fine-tuned to bring the numerical results into agreement with the actual observations (i.e., by comparing the model results against the actual observations of stars in an HR diagram). Modern stellar codes are very efficient, accurate, and fast, but for all that, they are still dependent upon the imposition of numerous physical simplifications. Be all this as it may. The point is, the equations of stellar structure can be numerically solved for on a computer, but solutions to these equations can be sought (that is, manufactured) to describe almost any evolutionary path that is desired for a specific model star.

It turns out, as we shall explore in more detail below, that to make a one solar mass model star evolve with a constant luminosity as it ages, one 'simply' has to impose a specific regime of internal mixing and mass loss. The problem now becomes one of reverse engineering. Given a specific set of requirements on how one wants a star to evolve as it ages, the appropriate internal/external manipulations (e.g., the extent of internal mixing and/or mass loss) can then be described. Indeed, we have already seen (in Chapter 3.3) the results of such manipulation at work. If one wishes to stop a star from becoming a red giant as it ages, then its interior must be well mixed — in other words, one must stop the build-up of a distinct composition discontinuity between the core, where energy is being generated and the composition is changing, and the rest of the star (recall Figure 3.2). While artificially invoking extra internal mixing will stop a model star from becoming a red giant, it still grows in luminosity as it ages. This can be countered, however, if one manipulates the mass of the model star at the same time. Specifically, by reducing the mass of a star as it ages, the luminosity increase can be controlled — even cancelled out entirely. While experimentation with computer models might tell us how to manipulate a star's future evolution (rendered as a horoscope in reverse), they do not tell us how such manipulation might be physically performed. As ever, the problem is in the physical doing rather than in the numerical tinkering. Here is our boundary of the unknown. At this stage, all that might be hoped for is that our far distant descendants — more removed from us than we are from our stone age ancestors, even our Burgess Shale proto ancestors — will have the intelligence and skill to engineer the Sun in a wise and informed manner.

4.1.1. *Hubert's laser*

To a great mind, nothing is little

(*A Study in Scarlet*)

If you want to engineer a prolonged main sequence lifetime, then you might try to induce extra mixing within a star's interior. This is illustrated in Equation (2.4) where the idea is to increase the f-term from its canonical value of 0.1 to a value closer to 1. In this manner, with $f = 1$, the entire star slowly changes composition, in a homogenous manner, rather than just the core. Even when fully mixed, the energy is still generated in the core because this is where the temperature and density have the values required to initiate fusion reactions. All that changes in this mixing scenario is that the entire hydrogen content of the star can be converted into helium within the core region as it gradually cycles through the interior. In the case of the Sun, the outer 1/3 by radius is convectively unstable and accordingly, well-mixed. This convective zone, however, does not extend to the inner core where energy is being generated. Accordingly, the inner 2/3 of the Sun by radius will need to be manipulated in order to satisfy the $f = 1$, homogeneous condition (Figure 4.1). Effectively, a pump of some kind will need to be engineered in order to both extend the star's main sequence lifetime and choke off its red giant expansion. The use of the word 'pump' here does not necessarily mean the construction of some form of mechanical apparatus to be placed within the star's interior, although that could conceivable be a possibility; rather, it is taken as a metaphor. Ideally what is wanted are the ways and means of making a star mix itself without the imposition of any internal mechanical systems.

There is a glaring fallacy that must be acknowledged in Figure 4.1. The diagram, as it is drawn and printed, is not wrong (per se), but it assumes that some technology will exist, at some time in the distant future, to do the job illustrated — that is, it will be technologically possible to mix the interior of the Sun at some far-off time. Just because I can draw a simple diagram, and even run a highly sophisticated computer program to back up the physical argument, this does not mean that such a process will ever be accessible to us, or indeed, any civilization billions of years in advance of our own. The human mind (remarkably) can actually imagine non-existent, non-physical, and completely impossible things. What this says about the human mind is certainly interesting, but it warns us that just because we can imagine something happening, it does not mean that nature can be manipulated and/or melded to perform the specific tasks being imagined. We can dream of making a star-mixing pump (our Figure 4.1), but if such technology is ever to be wielded, then it will have to be a very large and very subtle

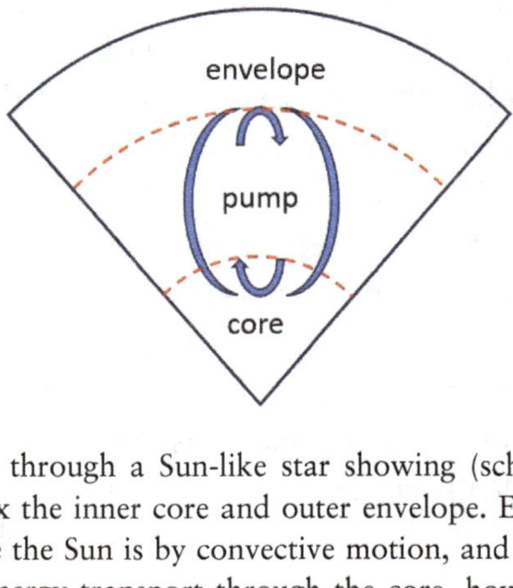

Figure 4.1. Segment through a Sun-like star showing (schematically) the pump action required to mix the inner core and outer envelope. Energy transport in the envelope of a star like the Sun is by convective motion, and accordingly, the envelope is well-mixed. Energy transport through the core, however, is via radiation, and the pump must act to mix this material with the envelope.

technology indeed. For all of the above, I am going to ignore my warnings. The problems are still there, they are still true filters, but I will attempt to resolve some of them in terms of non-miraculous — just massive and highly complex — actions.

French-Canadian astrophysicist Hubert Reeves first speculated on the origins of blue stragglers via externally induced mixing in his wonderfully rich book, *Atoms of Silence*, first published in 1981. Indeed, he argued that a powerful, external laser system might be used to generate a hot region within the deep interior of a star. The energy deposited by the laser would then induce convective turn-over, and accordingly, begin to mix the interior of the target star. This, of course, sounds reasonable enough, but the devil is in the details, and Reeves does, indeed, refrain from making any specific comments about the type of laser system required and/or the amount of energy needed to drive convective turn-over. No doubt, if one wanted to make the calculations, the system requirements could be determined. One can be sure, however, that a staggeringly powerful laser system would be required, and that certainly, the technical skill to perform such star-altering interventions do not, at present, exist on Earth. This, of course, does not mean that some highly advanced extraterrestrial civilization doesn't have the skill and technology to perform the task. It was through such reasoning that Hubert made the bold suggestion that perhaps blue stragglers are not only an odd class of stars, but are actually indicators of intelligent engineering being performed by extraterrestrial civilizations.

Now, recall, in the 1980s there were no clear theories about how blue stragglers might actually form, other than through some form of additional internal mixing within a star — the ideas of mergers within binary systems, mass transfer between binary companions, and stellar collisions had not, at that time, been clearly formulated. Accordingly, Hubert's suggestion that some form of external manipulation of a star might be going on, driven by a civilization of intelligent architects, is not unreasonable, and indeed, is rather an exciting idea. The argument is scientifically sound, and importantly, it sets bounds upon its viability. If Hubert's laser-mechanism is actually the process by which blue stragglers are made, then there is a clear prediction — all blue stragglers should show some evidence of a powerful laser in operation. The signal expected of such a powerful laser would be readily observable from Earth, and if present, would indicate that a highly un-natural process was in play — the conclusion being that extraterrestrial life exists within our galaxy. All good theories are testable and falsifiable (at least up to a point), and Hubert's suggestion is, by these standards, sound. At least two things have happened since Hubert made his argument, however. Firstly, natural mechanisms (that is via binary star evolution or stellar collisions) have been found for the origin of blue stragglers. Secondly, no one has reported the detection of any anomalous activity associated with a blue straggler that might betray the existence of laser light and/or a singular external energy source (see below). These two developments severely weaken Hubert's original hypotheses, but they do not fully negate it. Certainly, finding several 'natural' mechanisms for blue straggler production means that external manipulation is not an essential requirement for their appearance. And, certainly, finding no observational anomalies betraying external manipulation is informative. Accordingly, we can certainly conclude that not all blue stragglers are produced by external manipulation, but it is also the case that not all blue stragglers have been detected and/or observed at a level that fully rules out artificial formation.

Optical SETI (Search for Extraterrestrial Intelligence) is an extension of early SETI surveys which were developed to look for radio signals that might be beamed or leaked from extraterrestrial civilizations. Like all such surveys, optical SETI places its hopes in the hands of serendipity. The survey equipment scans for anomalous optical events from amongst a finite number of objects (those stars within the survey field) over a finite period of time. One hopes, accordingly, that the anomaly appears at the time that one is actually looking, and ideally, one hopes that it is not a signal that can be confused with some kind of a natural phenomenon. There is the additional caveat that one also hopes that any signal is not a unique, one-off event, and that it repeats itself over timescales of days, weeks, months and even years. Such repeat signals will allow for follow-up confirmation observations and detailed signal analysis. A unique signal is of no use to anyone, and indeed, this is something that has plagued the classic radio SETI detection known as the WOW event. In this case, a radio signal, with all the characteristics expected of a successful detection, was recorded on 15 August 1977. The signal endured for 72 seconds, and was recorded from the direction of the galactic center. Problematically, however, no repeat signal (in spite of overlapping field surveys conducted over many years) has ever been detected. To have a unique signal that cannot be explained away as a system defect is more frustrating than having no signal detections at all. Adding further frustration to the problem is the fact that the radio telescope that made the detection is no longer extant (the land that it once occupied was expanded into a golf course in 1998).

And, so it goes

(Kurt Vonnegut — *Slaughter House 5*)

A controversial, but natural explanation for the WOW signal has recently been offered by Antonio Paris and co-workers at the Center of Planetary Science [4]. In this case, the researchers note that the frequency at which the telescope was searching the sky was that corresponding to the hyperfine structure transition of hydrogen at a frequency of 1,420 MHz (or a wavelength of 21 cm), and this is exactly the frequency at which the hydrogen cloud surrounding an active cometary nucleus might be detected. If, Paris and co-workers argue, a visually undetected comet was in the direction of the telescope's beam, then it might have produced the WOW signal. Indeed, it transpires that two visually unrecognized comets were in the direction of the telescope's beam at the time the WOW signal occurred. This knowledge of past location is gained by back-tracking the orbits of periodic comets 266P/Christensen (discovered by Eric Christensen in 2006), and 335P/Gibbs (discovered by Alex Gibbs, in 2008). That active comets do produce vast hydrogen clouds is well known, and it is the natural result of the photodissociation of the water vapor molecules ejected from the sublimating surface of the comet's ice-rich nucleus. Indeed, the hydrogen cloud around a cometary nucleus can be many millions of kilometers in diameter, and importantly, it is known that they do radiate and are detectable with terrestrial radio telescopes working at a wavelength of 21 cm. The fact that the WOW signal was only recorded once is attributed to the motion of the comet, but, as pointed out by several researchers, this does not explain why the detection was not repeated on subsequent scans of the sky. This may be a quirk of the observations, and it reminds us that looking for the signals from alien civilizations is a non-trivial exercise.[3]

In the case of optical SETI, the hope is to detect some form of laser emission. Fortunately for such surveys, emission lines are rare within stellar spectra. There are some objects, certainly, that are defined according to the emission lines in their spectra (e.g., Be and Wolf–Rayet stars), but generally, one would not expect to see any emission lines at optical wavelengths, in say, a Sun-like star. Furthermore, a laser signal will be highly localized spatially and tightly tuned to just one frequency. A laser signal would stand out like a sore thumb. While many thousands of stars have been monitored in a number of optical SETI surveys, no signals consistent with an artificial laser have ever been detected. To my knowledge, no blue straggler has ever been specifically targeted in an optical SETI search, but as a group of odd stars, it might be useful to do so. The point here is that while, in general, the origin of blue stragglers can be explained perfectly well as a result of natural astrophysical causes, if an advanced — and it would have to be highly advanced — extraterrestrial civilization did decide to manipulate the properties of their planetary-system's parent star, then that star would potentially end up looking like a blue straggler. While we now know an awful lot about blue straggler formation, we certainly do not know everything, and the 'Other' channel in our Figure 3.1 may yet provide an avenue for profound discovery.

[3] While this book was in its proof-reading stage, Abel Méndez and co-workers, published a new re-analysis, of the "WOW" signal. Using archived data, they were able to refine the signals sky location (to a position slightly different to that given in previous studies). They also deduced a stronger signal strength, and a slightly different signal frequency to those that have been previously published. Méndez *et al.* conclude that the "WOW" signal was most probably related to a high radial velocity galactic source. Indeed, a signal generated by a maser flare situated within a small, cold, intermediate-velocity hydrogen (HI) cloud located within the galactic halo. See, https://arxiv.org/pdf/2508.10657.

4.1.2. *Criswell's star-lifting*

> *It is, I admit, mere imagination, but how often is imagination the mother of truth?*
>
> (*The Valley of Fear*)

While thoroughly mixing the interior a star will prolong its main sequence (that is hydrogen fusion) lifetime, it will not stop a star from becoming more luminous as it ages. For a habitable planetary system, this is problematic — indeed, it is the main reason why a civilization might wish to begin the external manipulation of its parent star. In short, as the central star increases in luminosity, for a given orbital radius, the temperature of a planet increases as well. Ultimately, the planet will become too hot for life to exist upon its surface. This fate will befall Earth in about one billion years from now, at a time still some four billion years before the end of the Sun's main sequence phase. To solve this biosphere heat-death problem, one can either attempt to enlarge Earth's orbit about the Sun, so as to combat the effect of its increasing energy output, or one could attempt to reduce the Sun's mass in order to reduce its energy output — one could also do both of these things simultaneously.

We saw in Chapter 1 that main sequence stars obey a luminosity-mass relationship of the form $L \sim M^{\alpha}$, where $\alpha = 3.5$ for Sun-like stars (recall figure 1.13). Accordingly, to reduce the luminosity of a main sequence star by, say, 1%, its mass must be reduced by about 0.3%. In the case of the Sun, this mass reduction amounts to some 6×10^{27} kg of material (or about 950 Earth masses). The Sun is presently losing mass, via the solar wind, at a rate of about 4×10^{16} kg/yr, and accordingly, if mass stripping by 0.3% is to be completed in, say, 10^{8} years, then this mass loss rate must be increased dramatically — indeed, by a factor of about 1,500.

There are a number of ways that one might attempt to increase the Sun's mass loss rate. One method would be to heat its surface — perhaps along the lines of a Hubert-like laser scenario, only this time not heating the deep interior to induce mixing. Alternatively, one might attempt to enhance the Sun's magnetic field, and through this effect, drive additional mass loss from the Sun's polar regions. David Criswell described just such a process in 1985, and argued that if one could induce a strong ring-current to circulate about the Sun's equator, then this would create a magnetic field capable of driving enhanced mass loss through two plasma jets directed along the Sun's spin axis (Figure 4.2)[4]. The process suggested by Criswell has become known as "star lifting" [2]. This name indicates that one is lifting matter from the outer layers of the Sun or, in the extraterrestrial

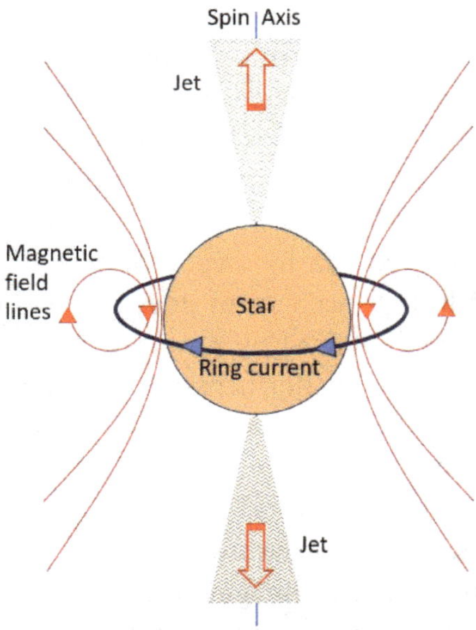

Figure 4.2. Schematic outline for the construction of a star-lifting mechanism. The key engineering step is to construct the 'ring current' girdle above the star's equator (blue ring). Once this is done ("ay, there's the rub") then it will produce a toroidal magnetic field (red lines) that will shape the star's outer atmosphere and drive matter outward (red arrows) via columnated jets orientated along the star's spin axis.

[4] As with Figure 4.1, there is the caveat to this simply-drawn diagram: just because we can imagine something happening does not mean that nature can be manipulated and/or melded to perform the specific tasks being imagined. For all that, the underlying physics of the star-lifting argument are perfectly sound.

civilization sense, parent star. The energy required to lift 1 kg of material from the surface of the Sun and move it to infinity is some 2×10^{11} Joules. The power P required to perform this lifting, at as mass-loss rate of $\dot{M}$, will be $P = 2 \times 10^{11}\ \dot{M}$ Joules per second. If we utilize the Sun's energy output to power the lifting, then $P = \alpha\ L_{\odot}$, where $0 < \alpha < 1$ is an efficiency factor. Accordingly, to achieve a 0.3% reduction in the Sun's mass over a 100-million-year time span, the mass-driving efficiency factor must be about $\alpha = 10^{-3}$ of the Sun's total energy output per second. The veracity of the star-lifting concept for controlling both the mass-loss rate and the evolutionary change in a star's luminosity was studied in detail by Matthew Scoggins and David Kipping (both of Columbia University) [5]. The numerical 'experiments' conducted by Scoggins and Kipping found that not only could the luminosity of the Sun be held constant through the controlled use of a star-lifting, mass-loss–driving mechanism, but that such manipulation would extend the main sequence lifetime of the engineered Sun by several billions of years[5].

Criswell's star-lifting mechanism does not require a miracle, or the imposition of some totally new physical mechanism, to work. It is predicated on basic physical principles, just scaled up to an incredible size of operation. Yes, the process works in principle, but will it ever be something that can be engineered by a galactic civilization — no matter how advanced? At this level of speculation, the answer is completely unknown. We can be stout-hearted and argue along the lines that if something can be imagined, then it can be built. Or, we can quibble endlessly about why engineering on the stellar scale is an impossibility for any civilization, and accordingly pay such imaginings no heed. In 1964, Nikolai Kardashev suggested a rather terse three-level measure of a civilization's technical abilities. A Kardashev

[5] Scoggins and Kipping coined the term *Lazarus stars* to describe such astro-engineered bodies. It is in the New Testament's Gospel of John that we are told how the deceased body of Lazurus of Betheny was resurrected and brought back to life. The star-lifting process, however, rather than resurrecting, or even reinvigorating a star, essentially acts to extend its main sequence lifetime only — in many ways they might be called *Tithonian stars*. In Greek mythology, it was Tithonus who was granted eternal life, but not eternal youth.

1 designation is applied to a civilization that can harness energy at a level comparable to that available from its parent star at the orbit of the host planet. To go further and become a Kardashev II civilization, the ability to harness the entire energy output of the planetary system's parent star is required. In the case of the solar system and our Sun, humanity will become a Kardashev I civilization once it can tap the entire 3.5×10^{17} J/s worth of solar energy intercepted by the Earth at its 1 AU orbital radius. The current energy consumption level for humanity is about 5×10^{13} J/s, so we are still sub-Kardashev I level by a factor of some 10,000 times. Our distant descendants will qualify as a Kardashev II civilizations once they can tap the full 4×10^{26} J/s worth of energy emitted per second by the Sun. The ultimate Kardashev III level is attributed to those civilizations that can tap the entire energy available from within their host galaxy. Such a civilization would have truly incredible power at its disposal, amounting to some 10^{37} J/s in the case of our Milky Way galaxy.

If our distant ancestors attempt to perform star-lifting on the Sun, in an effort to reduce its mass by about 0.3% over a 100-million-year time interval, then they will have to manipulate and control energy at a rate of about 4×10^{23} J/s. To do this, they will have to increase our present-day energy consumption rate by a factor of about 8 billion. It remains a moot point as to whether this level of advancement and engineering ability might ever occur on Earth, or anywhere else in the Universe for that matter. Arthur C. Clarke is remembered for his famous quip that "if an elderly but distinguished scientist says something is possible, he is almost certainly right, but if he says that it is impossible he is very probably wrong." Problematically, however, the question arises with respect to what we make of an elderly and undistinguished scientist who maintains that "it is impossible for Dr. Clarke to be correct."[6] One never wants

[6] In this case, the latter elderly scientist is certainly correct. The point being, that any assertion of an impossibility is equivalent to a statement of possibility. What Clarke is actually doing is simply assigning a very low probability to the impossible statement, and a very high probability to the possible statement. It is impossible for Clarke's two probability assignments to be both correct.

to say something is impossible, but it seems highly probable that there are no Kardashev III civilizations anywhere in the observable Universe, and it seems at least highly probable that there are no Kardashev II civilizations within the Milky Way galaxy. For all that, a civilization (orbiting a Sun-like star) with capabilities between that of a Kardashev I and II status might reasonably aspire to acquire star-lifting capabilities.

4.1.3. *Fogg's star exchange*

Tut, tut; we have solved some worse problems

(*The Priory School*)

We live in a throw-away society. Once an object is no longer fit for purpose, then we discard it and simply get another one. Pushing this (admittedly bad) philosophy to the largest scale, with a soupçon of recycling thrown-in for measure, Martyn Fogg has suggested that, perhaps, one can do the same with a planetary system's parent star [6]. If your parent star becomes too luminous, then swap it out for a less luminous one. The idea is grand, delightful, and outrageous, but not one requiring the occurrence of a miracle. Although, if one could engineer such an exchange, it would verge on the very edge of celestial magic. Fogg was inspired by the ideas introduced by David Criswell, and proceeded upon the basis that if one can engineer a Criswell magnetic engine about a star, then one can in principle control the field and mass-loss geometry to alter the space motion of the star. Hence the idea of stellar exchange. Of course, the idea would be fiendishly difficult to engineer (and this would be apart from even engineering a Criswell dynamo). The dynamics of the exchange would need to include both the ejection of the operating civilization's parent planet from the old star, and the capture of the same planet by the new one — at all times preserving life for all the residents. For all this, I don't doubt that one could compute an appropriate set of trajectories.

Fundamental to Fogg's star swap idea is the process of finding another star to set one's home planet about, and, indeed, this might be the most difficult part of the entire exercise. As we saw in

Chapter 1, in the galactic disc, stars (or more precisely, star systems) are invariably well-separated, being typically several parsecs apart, and if, as in our case, one lives on a planet in orbit about a G2 spectral type, luminosity class V star, then such stars are relatively few and far between. Only about 4% of the stars within the galaxy, recall, are Sun-like. By far the most common neighbor that any one star in the galaxy is likely to have, at any one epoch, is a low mass, red dwarf (spectral type M) star. In our case, at the present time, it is Proxima Centauri — an M5.5 V star, with a mass of 0.1221 $M_{\odot}$. Such a star still has a habitability zone, however, and there is no specific reason that an Earth-mass planet could not orbit about it. Problematically, in such an exchange, however, Proxima is an active flare star, and the new orbit would, if it was to provide the same insolation as the Sun at the present time, be so close-in that it would gradually slow Earth's spin period, driving it towards a synchronous condition with the spin rate being the same as the orbital rate (just as seen for our Moon). For all that, such low mass stars ($M < 0.25$ $M_{\odot}$) have a very long main sequence lifetime, and do not evolve into red giants. The list of pros and cons for such an exchange are near endless, but it is the idea, the mental picture, that really counts. The point being that if one truly did have the capacity to move stars and change stellar characteristics, what might one try to achieve, and how might such large-scale engineering be observable to us, on Earth, in the here and now? If one is attempting to move a star via some kind of Criswell dynamo, then certainly that star would display for observation an anomalously high mass loss rate, but it would be very difficult to observe and then infer any specific directional control over its spatial dynamics.

4.1.4. *Both shaken and stirred*

*There are in me the makings of a very fine loafer**

(*The Sign of Four* — * The term loafer is used here in its slang sense, and refers to someone who is a sluggard)

To mix the interior of a star, one has to perturb it, and one way of doing this is to fire a projectile, or projectiles, deep into its interior.

Here the idea is that as the projectile pushes into the interior of the target star, where it will deposit its kinetic energy of motion, and just like a stone thrown into a still pond, send out ripples and induce mixing between layers within the star. Once again, the idea is to make available, for central energy generation, as much of a star's hydrogen reserves as is possible. The mixing, as seen in earlier discussion, will extend the main sequence lifetime of the target star and quench its red giant phase. As with Hubert's laser, the energies involved are colossal, but if one allows for the eventual development of astero-engineering (see below), then there are plenty of potential projectiles that an advanced civilization might consider utilizing. Indeed, the projectiles could be Jovian planets, or even brown dwarfs. The number of rogue Jupiters sailing, unattached to any parent star, through the galaxy is, at present, only poorly constrained, but the number of planets within our galaxy is thought to be greater than the number of stars, and there are some 300 billion of the latter. The most common sub-stellar objects are the brown dwarfs, which have masses in the range from a few tens to about 100 times that of Jupiter. Indeed, in the same volume of space containing a single Sun-like star, it is expected that there may be some 23 M spectral type red dwarfs, perhaps some 5 brown dwarfs, and at least 20 planets with masses between that of Jupiter and Mercury.

By utilizing star-lifting technologies, such as the Criswell dynamo, a highly advanced civilization might engineer the mixing of a star (making it appear like a field blue straggler) by arranging for the collision between it and a single brown dwarf, or perhaps through multiple collisions with Jupiter-sized planets. Under these circumstances, one might hope to detect the destruction event (perhaps akin to that seen in V838 Mon — recall Chapter 3.11), and then, perhaps, a longer phase in which the target star is embedded within a tenuous gas envelope (akin to that seen in ZTF SLRN-2020– recall Chapter 3.11). On the basis that the target star was aged (that is, say, like our Sun, and near midway through its main sequence) prior to the impact(s), and fully mixed afterwards, then it would likely reveal, post-impact, chemical abundance anomalies within its spectrum. The mixing would result, for example, in a reduced surface hydrogen abundance, and an enhanced surface abundance of ^{3}He and ^{4}He (recall Figure 2.2).

4.1.5. *Looking for alien civilizations*

> *Life is infinitely stranger than anything which the mind of man could invent.*
>
> (*A Case of Identity*)

We do not know what engineering capabilities an alien civilization might be able to wield. Indeed, we have absolutely no evidence for any alien civilization existing anywhere, and any-when, in the entire Universe. At best, we might put our trusted and best scientific foot forward, and argue that no alien civilization can be founded on miracle-science or, indeed, science that defies basic physical principles as we know them. Such leg-work, however, hardly saves us from the marvelous actually coming about, since we (that is, humanity) do not know all there is to know about the underlying laws of physics within the Universe. This, of course, is not to say that many of the most basic physical principles are not understood, and we can make many strong predictions based upon what it is that we do know. The invocation of miracle-science (an oxymoron if ever there was one) will not help us, since on this basis absolutely anything goes, and accordingly any hypothesis, no matter how wild and improbable, about alien technology can be justified. This being said, we live in a Universe with, so to speak, time on its hands, and accordingly even extremely low probability events have a fair chance of eventually being realized, if not by now, then sooner or later in the future. Seeing the miraculous was one way that Arthur C. Clarke suggested we might be able to recognize a highly advanced civilization, but the problem is, what do you mean by miraculous, since we see miraculous things all the time? Photosynthesis is miraculous, the structure of DNA is miraculous, the existence of Earth is miraculous — indeed, to most people on Earth, the existence of computers and cell phones is miraculous (in that they have absolutely no idea how they work), and yet we do not consider these miracles as being the result of alien actions. The miracle that oversees the origin of life, that subtle spark of creation, is still an unknown, and one might argue, given the human penchant for corrupting and abusing knowledge, that this is a good thing. However, the one thing we do know about life is that when it does appear, it produces waste products, and it is this waste material that might enable us to find other civilizations. The late James Lovelock, who pioneered the idea of Earth as a self-regulating geo-biological system (Gaia), nailed this argument on its head when asked, in the 1970s, how we might recognize (from Earth) the existence of present-day life on Mars. His argument was short, straightforward, and deep-sighted. What you look for, Lovelock reasoned, is an unexpected atmospheric chemistry, and the presence of elements that should not be there. That is, in the case of the Earth, the atmosphere is largely composed of nitrogen and oxygen, in complete contrast to those atmospheres surrounding Venus and Mars. Likewise, Earth's atmosphere shows a distinct methane component, a component that requires continual maintenance. If there were no continually acting set of agents involved in maintaining the Earth's atmosphere, then it would not appear as it is observed to be today. The key agent in explaining the Earth's present atmosphere is life. Specifically, the level of oxygen in Earth's atmosphere is entirely the result of the evolution of life and its continued and near-constant levels throughout the age of the Earth is a direct result of life interacting with and moderating long-term geological feedback processes (this is effectively Lovelock's Gaia hypothesis). The fact that the atmosphere of Mars is in stable chemical equilibrium and predominantly composed of CO_2 is very strong evidence that there is no present-day life on Mars. This, of course, is not to say that life didn't exist on Mars in the distant past. An alien civilization with the capability of monitoring (through spectroscopy) Earth's atmosphere will have noticed, if their records trace back just a few hundreds of years, the emergence of us (that is modern techno-humans) through the very rapid increase in atmospheric CO_2 levels, and the sudden appearance of numerous manufactured pollutants. Even if their records traced as far back as a few thousands of years, they would still find the sudden and rapid growth of CO_2 in Earth's atmosphere anomalous. That anomaly on Earth, of course, is the product of anthropic actions, industrial pollution, and life-style choices. It is entirely possible that the observation of such chemical anomalies in exoplanet atmospheres will reveal the miraculous. Even this argument of recognition through anomaly, however, has its limitations, since, one might hope (or even reason) that a highly advanced civilization will be smarter than past generations of human beings have been, and not pollute and despoil their home planet. This all points to the idea of there being distinctly recognisable epochs in the development of a planetary civilization. If an advanced alien civilization scanned the Earth's atmosphere a billion years ago, they would have found evidence of biotic, oxygen-generating life, but no evidence of advanced life, and indeed, might have reasonably deduced that there was little prospect for advanced life ever appearing — and then a miracle, by which I mean a very low-probability event, happened. On this note, Adam Stevens and co-workers [7] have noted the irony that one of the most distinctive biosignatures that would indicate the detection of an alien civilization might be through the short-lived appearance of methanethiol (CH_3SH) in the spectrum of an exoplanet's atmosphere. This detection would betray the appearance of a very distinct molecule associated with biomass decay and death. This detection would indicate a recent, mass die-off. Other distinctive biomarkers are also available to researchers, but these can often be generated by natural non-biotic processes. A recent example of this concerns the potential detection of phosphine in the upper cloud deck of planet Venus in our Solar System. Certainly, phosphine is a biproduct of anaerobic ecosystems on Earth, but it can also be produced abiotically, through the weathering of volcanic rocks. For this biomarker, context is everything. Its detection in the Cytherean upper atmosphere is particularly interesting, since there is no clear indication that it is derived through ongoing volcanic activity [8].

It is entirely conceivable that alien civilizations, literally billions of years more advanced than our own, exist within the Universe — even within our galaxy. Indeed, the Sun is a relatively young star, with most of the other stars in its galactic neighborhood being of order a billion years older. The problem, of course, is how do we recognize such advanced civilizations since, in a time and evolutionary sense, they will be as far removed from us as we are from cyanobacteria. One additionally assumes that this recognition problem works both ways, and that any highly advanced alien civilization will actually recognize human beings as a being a knowing and knowledgeable creature. For all that, apart from some revelatory appearance, we can only search for such advanced civilizations with technologies that we understand. Accordingly, the search is directed by serendipity, and the chance detection of stray (possibly beamed) electromagnetic radiation emissions (this approach includes, optical SETI, and the now traditional radio SETI), or through attempts to alter a star with very-high-powered laser or through anomalously high mass-loss rates engineered by star-lifting activity.

One way to maximize detection possibilities would be to look at planetary systems in orbit about a star that is well on its way towards ending its main sequence phase. Figure 4.3 shows a diagram of age versus mass for the subset of stars known to be supporting planetary systems. The star masses deduced from the observations range from about 0.1 to 2 $M_{\odot}$ (the Sun is shown in the diagram by a white asterisk). The star ages range from a few millions of years to some 10 billion years. Also shown in Figure 4.3 is the diagonal line representing the mass versus main sequence lifetime relationship. Any star plotting above this line has reached the end of its main sequence lifetime and is headed towards becoming a red giant. It is amongst these stars, and those stars close to the main sequence limit, that one might expect to see some form of stellar engineering taking place. Literally, for the planets in orbit about these stars, time is running out, and a rapid and extensive change in each system's habitability zone is about to take place. When the Sun becomes a red giant, in about 5 billion years, it will swell to a size that will consume planets Mercury and Venus (and possibly Earth), and its luminosity will increase briefly by a factor of several thousand, and then, for a longer-term during core helium burning, to about 50 times its present output. Life, at least as we know it on Earth now, cannot survive beyond the Sun's main sequence lifetime [2]. If our distant descendants want to survive the deadly, end–main-sequence

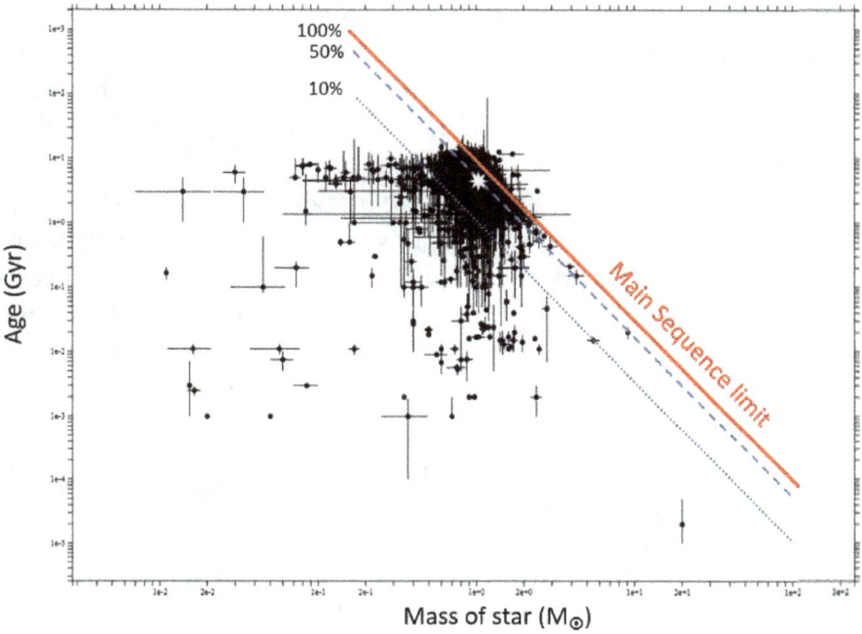

Figure 4.3. Age versus Mass of parent stars for exoplanetary systems. Each of the stars plotted in this diagram hosts one or more planets. The red diagonal line indicates the main sequence lifetime for stars of a given mass (recall Equation 2.4) — the blue dashed and the blue dotted diagonal lines indicate ages equivalent to 50% and 10% of the main sequence lifetime for a given star mass. The Sun is indicated by a white asterisk. Background image from the Catalog of Exoplanets — https://exoplanet.eu/catalog/.

transition, they must either abandon the solar system to become space-faring wanderers, or attempt to modify the Sun. The long and persistent existence of the Fermi Paradox[7] suggests that few, if any,

[7] Enrico Fermi asked, in 1950, if advanced alien civilizations exist within our galaxy, why it is that we have no sign of their ever having visited our solar system, to either colonize it or to stock-up on resources. Even a modest expansion rate would allow for an advanced alien civilization, with space-faring technology, to spread throughout the entire galaxy within a few millions of years. That there is no indication of the Earth, or, indeed, the solar system at large, ever having received a visit from a space-faring colony is, in fact, powerful evidence to suggest that galactic travelers do not actually exist. Indeed, we are bound to conclude that large-scale interstellar space travel has never been achieved or perfected within (at least) our galaxy.

advanced civilizations have attempted to leave their planetary systems at the terminal main sequence phase. Accordingly, if they are to survive, then they must set about the process of engineering their parent star into a more life-compliant, life-extending state. Well, this or totally perish.

Taking the nominal main sequence lifetime, τ_{ms}, of its parent star to be the absolute time limit that any non-engineering civilization might survive, then an advanced civilization must begin the engineering of its parent star prior to a time of $\beta\tau_{ms}$, where $0 < \beta \leq 1$. In our case, the solar system has a beta-factor of $\beta = 0.45$, and this indicates that many tens of millions of years can pass before we need worry about engineering the Sun. Star and planetary systems plotting near the main sequence lifetime limit diagonal in Figure 4.2, however, have beta-factors of order $\beta \approx 1$, and accordingly, if any of these systems support life, in any form that might possibly resemble ours, then they are literally at the brink of an existential catastrophe. It is in these planetary systems, if anywhere, that we might hope to find evidence for star engineering of some form [9].

Many suggestions have been made concerning what an advanced extraterrestrial civilization might attempt to do with respect to its parent planetary system. We have so far discussed engineering of the parent star, but another option, prior to the $\beta \approx 1$ imperative, might be to tap the entire energy output of the host star. Indeed, such an action would raise a civilization to a Kardashev II level. Freeman Dyson suggested, in a classic 1960 research paper, that one way in which this total energy trap might be achieved is to construct an artificial sphere[8] around the star. In this manner, the entire energy output of the star can be put to work with respect to the host

[8] Detailed calculations (actually basic physics and material science) indicate that a spherical shell about a star will not be stable against small displacement perturbations, so a Dyson sphere would really have to be a spherical swarm of islets, each capable of moving in unison but isolated from its nearest neighbors — like a murmuration of starlings.

civilization[9]. Even though a Dyson sphere taps the entire energy output from its central star, the work done with all that energy must eventually be reradiated into space. This will make such objects strong infrared sources. Not only this, however, a Dyson sphere would radiate as a near perfect blackbody radiator (recall Chapter 1). This latter feature is something that is rarely achieved in nature[10], and accordingly, any strong infrared source with a perfect blackbody spectrum is a candidate for being a Dyson sphere. Various searches have been made for such structures, and while numerous candidate systems have been found in survey data, none to date has proved fully convincing of being an actual Dyson sphere [10].

It is not necessarily the case that an advanced civilization will first construct a Dyson sphere before moving on to consider star-engineering. Indeed, once a civilization has the capability of building a Dyson sphere around its parent star, they would also have the technical capability to begin star-lifting, thereby extending the parent star's main sequence lifetime. The latter option is potentially the better one to adopt, since it extends the lifetime of a planetary system's habitability zone. In other words, by attempting to reset the main sequence lifetime of a host star (that is, make it a blue straggler), one potentially lengthens the time over which a civilization can exist within its planetary system. By building a Dyson sphere, a civilization essentially limits its habitability zone to that of the outer surface of the sphere. Furthermore, a Dyson sphere will become uninhabitable once the parent star evolves away from the main sequence, if not through the physical expansion of the star, then through its tremendous increase in luminosity. On this basis, one

[9] There may be a fundamental flaw in Dyson's basic premise of energy maximization, and that is the drive, in fact, to make things more energy efficient. Rather than building 'big' to tap more energy from the Sun, it might be more productive to develop cooler-operating, very low-power systems.

[10] The only situation in which a near perfect blackbody spectrum has been recorded is that of the cosmic microwave background — the imprint of the Big Bang itself. And, we are fortunate indeed, that while being very close to a perfect blackbody radiator, there are, nonetheless, very small deviations. It is from these ultra-small deviations that the large-scale structure of the Universe (and Earth and us) eventually evolved.

might reasonably argue that star-lifting is a better, long-term option for an advanced civilization than building a Dyson sphere.

There is an elephant in the room. And that elephant is us. Life evolved very shortly after Earth formed; indeed, some evidence suggests that bacterial life was present within a few hundred million years of the final large accretion impacts. The machinations of evolution have resulted in our appearance on planet Earth within less that half the main sequence lifetime of the Sun. And, while we are not a highly advanced galactic civilization (to the level being discussed in this chapter), we may be the early forerunners of such a civilization (or we may not). Indeed, tossing the half-brick of contempt at the Copernican Principle, humanity has made its appearance well before the Sun is even halfway through its main sequence phase. We are, for so it seems, especial early-arrivers in the advanced-lifeform game [11]. Indeed, the likelihood of advanced life (meaning something like an oxygen-breathing mammal capable of using and manipulating high-level technologies in this context) coming about, even when all the basic ingredients are available, is incredibly small, and it is not unreasonable to suggest that the most likely time at which an advanced civilization, capable of conducting advanced space engineering, appears, T_{rise}, will be comparable to the main sequence lifetime of the host system's parent star: that is, $T_{rise} \sim \tau_{ms}$. Under these circumstances, an advanced civilization is only likely to appear, in a given habitable planetary system, at the very moment when it is effectively too late to do much about stopping the red giant phase of the parent star from coming about and destroying the newly emerged civilization. By this reasoning, there is likely not enough time for a given civilization to effectively develop its engineering prowess to begin building either a Dyson sphere, or enter into a star-lifting campaign, before it is destroyed. Well, such may be the doom of most alien civilizations, but we are the foil to that argument. And our descendants have many hundreds of millions of years yet to develop their potential star-altering skills. There are reasonable counterarguments to such 'hard steps', late-arrival scenarios, and Daniel Mills (Ludwig-Maximilians-Universität, München) and co-workers have recently suggested that our emergence time is governed and controlled according to planetary and geological timescales [12]. For all that, the equality $T_{rise} \sim \tau_{ms}$ is still likely to be the typical outcome within any habitable planetary system[11]. Such conditions actually provide us with a somewhat weak answer to the Fermi Paradox. It is only a paradox on the basis of Fermi's assumption that there are, and have always been, many advanced alien civilizations within our galaxy, and we do not know if this is actually true. Given the possibility that advanced civilizations are going to be, at best, very, very rare at any given galactic epoch, then their apparent absences from our solar system is readily explained, and indeed, entirely to be expected.

For all of the above, important as it is, in this chapter we are in the realm of creative imagination. There are no real answers available, as yet, to any of our speculations concerning the future of life on Earth[12], the future skill levels of our most distant descendants, and/or their potential ability to construct such structures as a Dyson sphere or begin the act of star-lifting. Likewise, there are no real answers available about the past and/or present existence of advanced civilizations within our galaxy. All that can be said at the present time is that there is absolutely no observational evidence to support the existence of extraterrestrial civilizations. For my part, I am certainly prepared to accept as highly probable that advanced civilizations do exist within our galaxy, and that they have and can manipulate their local planetary systems through terraforming and possibly even astero-engineering. But, for all that, I do not think that effective, large-scale interstellar space travel has ever been (or will

[11] If nothing else, it certainly sets an upper limit to the emergence time.

[12] Here we should remind ourselves that, even if we survive the next several centuries, it will not be human beings (*Homo sapiens*) that develop the space-faring skills and engineering prowess to alter the Sun. Rather, these descendants will be as far removed (and much more) from us as we are from our Australopithecus ancestors.

ever be) perfected (or indeed, is really even a sensible action[13]). Continued observations, the ultimate selection effect, will help resolve some of the issues (eventually — perhaps), but I would wager good money that these same observations will provoke more questions than provide actual answers.

4.2. Astero-engineering

By profession I am a hydraulic engineer

(*The Engineer's Thumb*)

The idea that a star might be artificially manipulated is both bold and audacious [13]. It is an entirely unreasonable concept, and certainly it is an action well beyond current human capabilities. For all that, we do know a good deal about the properties of stars, and what physics takes place within their interiors. Up to this point we have only considered the extended mixing of the interior of a star, or the reduction of its mass, as a means of extending the main sequence lifetime, but there are other possibilities. What stellar modeling tells us is that the main sequence lifetime of a star can be extended by manipulating any, a selection of, or all of the following properties:

- Mass loss
- Internal mixing
- Non-thermal pressure support
- Spin rate
- The radiation flow through the interior

[13]The technology for interstellar space travel would certainly be in reach for any civilization capable of building a Dyson sphere or performing star-lifting, but I do not easily accept the general precept that the incentive to pursue large-scale, many-bodied interstellar space travel really exists — and this is so for many reasons. Indeed, if the incentive to become interstellar gypsies really does exist, then the existence of the Fermi Paradox suggests that it is a doomed and fruitless exercise from which very few (zero?) practitioners have survived. For all that, I would be more than happy to be totally wrong on this point.

- Radiation trapping at the surface
- Central black hole seeding and accretion

As already indicated in earlier sections, reducing the mass of a star lowers its luminosity and thereby extends its main sequence phase. Mixing material from the outer envelope of a star to its inner core also extends its main sequence lifetime and, in the fully mixed state, stops its red giant phase from coming about. Furthermore, by introducing some non-thermal pressure support (e.g., through the generation of a strong magnetic field) at the center of a star, its core temperature is lowered, the energy generation requirements are reduced, and the main sequence lifetime is extended. By spinning up a star, one can potentially induce extra mixing within its interior and also provide some non-thermal pressure support — both processes tending to increase the main sequence lifetime. The opacity term expresses the difficulty with which radiation passes through the interior of a star, and while it is a complex function relating to the composition, temperature, and density at each location within a star, it could, in principle be manipulated (by the addition of artificial additives) to reduce a star's luminosity. The entire structure of a star could, in principle, be manipulated by placing a reflective wrap around its surface (see below), allowing, thereby, the temperature of its outer layers to increase. Alternatively, one might attempt to take control of the central energy generation mechanism, and rather than letting nuclear fusion reactions power a star, one might power it through the accretion of matter on and into a centrally located black hole [14, 15].

There are many ways, in principle, to alter what a large collection of matter actually does — the problem, of course, is having the engineering skill to do it. The physics is not so much that of the miraculous, nor is it dependent upon the discovery of entirely new physical principles; the essential issues is one of the engineering scales, and the extremely harsh environmental conditions — conquer these, however, and you can control what a star does. Such actions are challenges set for the distant future inhabitants of Earth, and these feats may possibly be something already achieved by ancient and advanced civilizations elsewhere within our galaxy and others.

4.3. A solar wrap

I confess that I have been blind as a mole, but it is better to learn wisdom late, than never learn it at all.

(*The Man with the Twisted Lip*)

If you place an electric lightbulb inside of tin with a perfectly reflecting interior, close the lid and turn on the power, what will happen? While this is not an experiment to try at home, the end result will be the destruction of the lightbulb. At first, the energy radiated by the lightbulb will heat the air inside of the tin. Since this is an idealized tin, with perfectly reflecting sides, no energy is actually absorbed by the tin itself, and it remains cool to the touch (in real life, the tin will most definitely become warm to the touch). As time goes by, however, the air inside of the tin will get hotter and hotter as more and more energy is pumped into it by the hot lightbulb. Indeed, the air will eventually reach a temperature at which the material that makes up the lightbulb must fail — the solder holding it together will melt, or the glass cover sealant will begin to burn. Either way, the end result will be destructive, and the energy source (the lightbulb) inside of the tin will be destroyed. Now, what happens if you wrap a perfectly reflecting foil around the outer surface of a star? Will the star become ever hotter and hotter, or will it do something else? The answer to this question, perhaps surprisingly, is that it will do something else. But what?

We are not talking about a Dyson sphere in this thought experiment, since the idea behind such a sphere is to collect the energy radiated by the star on the inside, use that energy for various purposes, and then re-radiate the waste energy, at infrared wavelengths, into space from the outer surface of the sphere. No. In this case, the idea is that the foil wrap is perfectly reflecting, with no energy being radiated into the surrounding space [16]. In this manner, the outer layers of our wrapped star will heat up and expand somewhat (it is a stretchable wrap). As time goes by, the outer layers of the star will become warmer and warmer. Ultimately, the star will become isothermal, with a constant temperature throughout its interior — that is, a uniformly hot plasma. The energy demands of the star now no longer require fusion reactions to take place in the central core. Recall from Chapter 2 that a star only needs a central energy source to stay warm within its interior to fend off gravitational contraction and replenish the energy radiated into space at its surface. Now, since no energy is radiated into space at its surface, the temperature of the isothermal interior will adjust itself to lie just below that at which fusion reactions switch off. In effect, the star evolves into a uniformly hot, quivering gas. Provided these conditions hold, the wrapped star becomes a vast foil-wrapped sphere of hot gas stable against gravitational collapse indefinitely. Unlike the light bulb in a tin, a star that has been wrapped in a perfectly reflecting foil will not heat-up until it explodes, but it will actually cool down overall to a uniform temperature. In effect, the star blinks out. Since the warp is perfectly reflecting on the inside (this is a thought experiment, remember) it will not radiate any energy into space, and accordingly, it will be completely undetectable to any telescope, no matter what wavelength of the electromagnetic spectrum the telescope is working at.

Well, the devil, even in a thought experiment, is in the details. The only engineering issue, however, is that of developing a perfectly reflecting cover, and then making enough of it to wrap a star. No such perfectly reflecting material is known at the present time — there is always some absorption, so one is clearly looking at producing some form of carefully manufactured metamaterial. For all that, there is still a practical side to the star wrap idea. Although no energy is radiated from the outer layer of the wrap, one could still tap energy from the interior to power a civilization situated on or close to the wrap's exterior. In some sense, a star wrap is the ultimate blue straggler. It has the mass of a star, it has a fully mixed interior, and it is not evolving at all. If one taps energy at the surface, however, the gas inside of the wrap will gradually cool, and ultimately it is possible that a so-called gravithermal collapse will take place. The drive now, however, for a civilization living off the energy being extracted from the interior of a wrapped star is to conserve as much energy as possible. It must strive to live in its lowest possible energy state, extracting as little energy as is possible from the interior of the wrapped star. In this manner a civilization could live almost indefinitely. These characteristics are almost the exact opposite to those looked for with respect to a Dyson sphere and/or a Kardashev I civilization. Perhaps the long-silence of SETI is exactly what we are supposed to see and hear, and the Fermi Paradox is solved by the very fact that ancient, technologically advanced civilizations are comfortable, thank you very much, tapping as little energy as possible from their wrapped, entirely invisible parent stars.

Can stars ever entirely blink out and disappear from sky-survey images? Interestingly, the answer to this question is yes, some stars do indeed seem to literally disappear from view (Figure 4.4). Now,

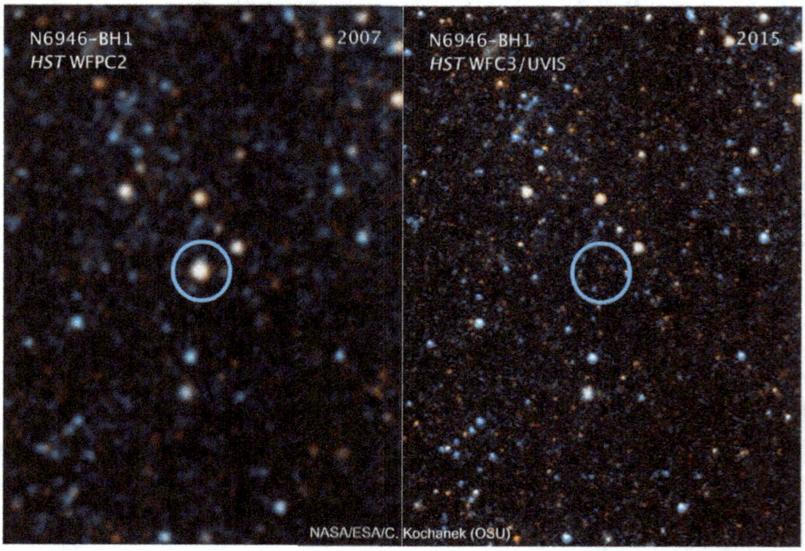

Figure 4.4. Two Hubble Space Telescope images of the supergiant star N6946-BH1 and its surroundings, taken 8 years apart. At some point in the time interval between 2007 and 2015, this massive star (weighing in at some 25 times the mass of our Sun) appears to have collapsed directly in to a black hole without any significant ejection of matter into surrounding space. Image courtesy of NASA, ESA, Hubble, C. Kochanek (OSU) [18].

there are many reasons why such a disappearance might take place: we could be observing a variable star, which at its faintest limit passes below the detection threshold of the camera being used; we could be observing a star in an eclipsing binary system, but the disappearance in such cases would only last for a number of hours; we could be looking at an R Coronae Borealis-type variable, which semi-periodically dims because of the generation of an expanding, cool, carbon-rich dust envelope; we could be looking at a flare star, bright at the time of detection, but below detection threshold during times outside of flare generation. We might also be seeing something highly exotic. Perhaps a star in its completion phase of being wrapped; perhaps a failed supernova[14]. In order to test the idea that stars might mysteriously disappear from view, Beatriz Villarroel (Nordic Institute for Theoretical Physics, Stockholm) established the Vanishing and Appearing Sources During a Century of Observations (VASCO) catalog [17]. This catalog is based upon data extracted from numerous star survey catalogs constructed over the past 70, or so, years — from old photographic survey plates to present day digital imaging. Careful digital processing and cross-comparisons between survey positional data has yielded evidence for some 100 stars (out of nearly 100 million surveyed) completely disappearing. It is important to emphasise that the stars that apparently disappeared from view did not do so because of having a high proper motion (recall Figure 1.4), or because of being some image defect, or by going nova, or by any other known star-variability mechanism. They simple seem to have vanished from view at some time within the 70-year time interval between comparison surveys. It is probable that there are straightforward, natural explanations for each of the objects in the VASCO catalog, but it is additionally possible that they hint at more mysterious processes, possibly astero-engineering, taking place within our galaxy. As with all such cosmic mysteries, we are left to wonder and speculate, our thoughts hovering on the quicksilver boundary between what we think we know and what we

[14] These are conjectured to be massive stars that collapse directly into a black hole state, without undergoing a supernova outburst phase.

plainly do not. Such thoughts only bounded by the rejection of miracles, and only partially tempered by the point of circumspection: *Omne ignotum pro magnifico*[15].

4.4. Swallowing a black hole

It is a possibility and we cannot afford to disregard it.

(*The Naval Treaty*)

When the Large Hadron Collider (LHC) at CERN was due to be commissioned in late 2008, it came under a Hawaiian court action that questioned its safety. At stake, the plaintiffs claimed, was the very existence of the Earth. The lawsuit was simple-minded and based upon incorrect physical reasoning, but it did tap into an existential fear predicated upon the idea that the LHC might produce, from within its colliding proton beams, a multitude of mini-black holes [19]. What if, it was reasoned, all these low-mass black holes produced by the LHC sank to the center of the Earth, and there coalesced to form an ever-larger black hole? Surely, once massive enough, the black hole would start to accrete more and more material from the Earth's interior, hollowing out our terrestrial home, eventually drawing it and us into its crushing singularity. The court case was ultimately thrown out (indeed, it should never have moved forward) since it is patently clear that the concern of such deadly destruction is entirely unfounded. The reason for this rejection, and the certainly of such rejection, is actually expressed by our very existence, and indeed, the ponderous weight of the 4.56-billion-year history of Earth's existence. While the LHC does enable the most energetic collisions between protons ever engineered by humankind, the cosmos has been performing similar such experiments, with cosmic rays, at much higher energies in Earth's atmosphere for the entire history of our planet. If mini-black holes can be produced by

[15] From the Roman writer and historian Cornelius Tacitus: "everything unknown appears magnificent."

the collision of atomic particles, then the Earth would have been destroyed billions of years ago.

If, for the sake of argument, however, the LHC was to generated copious numbers of mini black holes (and there is absolutely no evidence that it has produced even one) then they would still be totally harmless entities. Indeed, the most massive mini-black hole that the LHC might be able to theoretically generate would have a mass in the order of 10^{-23} kg (based upon a collision energy of some 10 TeV = 1.6×10^{-6} Joules). This mass is incredibly small, and it is, in fact, much smaller, by some 15 orders of magnitude, than the smallest-mass black hole that it is thought possible to exist[16]. In order to have a size[17] at least as large, say, as a hydrogen atom (some 10^{-10} m across[18]), a black hole must have a mass in the order of 3×10^{16} kg (this is equivalent to the mass of a 30-km diameter stony asteroid). Even a black hole the size of a hydrogen atom, however, would struggle to accrete matter with any great efficiency (but see later). The LHC posses absolutely no threat to our continued existence. In terms of astero-engineering usage, however, it is this latter issue that is actually important, since, in principle, a star can be powered by accretion of material onto a centrally-located black hole. The point is, however, that the black hole has to be large enough to enable the efficient accretion of material over an extended period of time. If these two issues of accretion rate and time can be accommodated, so that the lifetime against destruction is actually longer than that of its erstwhile main sequence lifetime, then there is, in fact, an advantage to letting a centrally located black hole feed upon its surrounding stellar envelope.

Shortly after introducing the idea of low-mass, primordial black holes (PBH), Stephen Hawking suggested, in 1971, that such a PBH might be located at the center of the Sun [20]. At that time, the notion was to partly power the Sun through central accretion, and

---

[16] The smallest possible black hole mass is that having an event horizon equal to the Planck length (some 10^{-35}-m), and this corresponds to a mass of 2×10^{-8}-kg.

[17] By this we mean the diameter of the black hole's event horizon.

[18] The 'classic' Bohr radius of the hydrogen atom is some 5×10^{-11} m.

thereby reduce its central temperature and the rate at which the hydrogen fusion reactions were running. Indeed, the idea was to solve the so-called missing neutrino problem. In the early-1970s, the observations indicated that the solar neutrino count was about 1/3 of that predicted by state-of-the-art stellar models, and the question was why. It is now understood that the answer lies in the properties of the neutrino and its ability to oscillate between three flavors. Since the neutrino detector was only sensitive to one neutrino flavor, we now know that the 1/3 count was actually correct, and that no adjustment is required of the hydrogen fusion rate at the Sun's core. For all that, at the time, Hawking's suggestion was perfectly sound — by reducing the temperature at the center of a Sun, one reduces the level of energy generation through fusion reactions, and, in the process, the number of neutrinos produced is also reduced (recall Figure 2.2).

The idea that a star might be powered by accretion onto a centrally-located, compact object was, as we saw in Chapter 3, explored by Kip Thorne and Anna Żytcow in the late-1970s [21], and the idea can be extended to a general astero-engineering principle. Clearly, however, if this engineering approach is going to be performed, one must first find a PBH, and currently there is zero evidence for their existence — which, of course, does not mean that PBHs do not exist, but simply that the definitive observational evidence for their existence hasn't been identified to date.

In terms of astero-engineering, what sort of initial-mass black hole are we talking about? Firstly, on the basis that it is a PBH, we have to allow for the finite lifetime of such objects. Once again, it was Stephen Hawking who introduced the idea that black holes are not permanent structures, and that they slowly 'evaporate' away. The evaporation lifetime T_{evap} against Hawking radiation is related to the mass of the black hole and, $T_{evap} = 10^{-23} M^3$ years, where the mass is expressed in kilograms. Taking the Universe to be some 13.8 billion years old, the smallest PBH that might exist at the present epoch has a mass in the order of 10^{11} kg, and at this mass it will have a surrounding event horizon smaller than a single hydrogen atom in size. For the central black hole accretion scenario to play out, therefore, one must assume that there is a whole range of low-mass black hole candidates available, with masses well above the theoretical minimum for a PBH, capture, manipulation, transportation, and eventual insertion into a host star.

Never missing the opportunity to run a computer simulation, however, many researchers have considered how a star being powered by a centrally accreting black hole might appear and evolve. The models of such Hawking Stars, as they have recently become known, must be highly parametrised, but they can (with no surprise) be made to 'work' [22]. The key change from a traditional stellar model is the way in which such bodies generate energy within their interiors. Rather than generating energy at their centers entirely through hydrogen fusion reactions, Hawking Stars are powered by accretion onto their central black hole. In this way, material falling towards the black hole's event horizon begins to pile-up, and accordingly, becomes compressed and heated. The energy generated per second, L, in this process is related to the accretion rate, $\dot{m}$, such that $L = \varepsilon \dot{m} c^2$, where c is the speed of light and $0 < \varepsilon < 1$ is a radiative efficiency term. The luminosity can also be approximated according to the so-called Eddington luminosity, L_E, named after pioneering astrophysicist Sir Arthur Eddington. L_E is related to the mass of the black hole M_{BH}, such that $L_E = 4\pi(c/k)\, GM_{BH}$, where κ is the opacity, and G is the universal gravitational constant. At the high temperatures expected at the accretion boundary, the opacity term κ, dominated by so-called electron scattering, will generally be a constant term related to the hydrogen abundance. The mass of the black hole will increase as it feeds upon the host star's matter at an exponential rate: $M_{BH}(t) = M_{BH}(0) \exp(t / \tau)$, where t is the system age, τ is a growth scaling constant, and $M_{BH}(0)$ is the initial black hole mass. The art of modeling a Hawking Star is to parameterize the initial black hole mass and the accretion efficiency. The central black hole mustn't be too large, since then it would rapidly consume its host, and not too small in that it has no specific effect on the duration of the main sequence lifetime. Most computer modeling scenarios assume an initial black hole mass of $M_{BH}(0) \approx 10^{20}$ kg, this being an object with an event horizon of radius about 1.5×10^{-7} m $\approx 2800\, a_0$, where a_0 is the classical Bohr radius of the hydrogen atom in its ground state.

According to the value of $M_{BH}(0)$ and the growth scaling constant τ, the type of object a Hawking Star might begin to emulate is, in fact, a blue straggler. This will be the case once the accretion luminosity begins to outcompete the energy generated by hydrogen fusion reactions during the host star's main sequence phase. If, on the other hand, the accretion rate is sufficiently slowed, so that the host star has migrated away from the main sequence, then it will begin to emulate what might be called a red straggler: a bloated, low surface temperature star, with a black hole heart, denied the opportunity of igniting helium fusion reactions within its core. Nothing is ever easy in the world of astero-enginering, however. Indeed, for a Hawking Star to even form, a special set of circumstances must come about. In contrast to the situation of Thorne–Żytcow stars, where the accreting neutron star component is taken to have spiralled inwards to the center of the host star, it is much more difficult for a star to capture a primordial black hole. Rather than settling to the center of a host star, a PBH is more likely to pass right through it. We would, accordingly, expect Hawking Stars to be an extremely rare natural phenomenon, even if PBHs are very common (as some recent dark matter theories intone). For all that, a Hawking Star might conceivably be engineered by a suitably advanced civilization. The task, of course, would be incredibly complicated, dealing with extreme sizes (both extremely small and extremely large), temperature domains, and dynamics. Provided one can find and capture a 10^{20}-kg PBH (recall an object just a few thousand times larger than a single hydrogen atom), the astero-engineer would then need to place this object at the center of a host star. Simply dropping the PBH into the host star will not necessarily result in it simply settling to the centre[19], so some form of damping apparatus would be needed

[19] If one simply released the black hole from rest at the surface of the Sun, it would oscillate across the Sun's diameter, back and forth, like the manic pendulum described, long ago, by Edgar Alan Poe. The period of oscillation will be about 2 hours and 47 minutes.

to carry the PBH to its eventual location in the host star's core. All this is arguably achievable (at least in a thought experiment) by a civilization sufficiently advanced in material sciences, space mining, PBH detection and capture, and transportation control capabilities.

4.5. The uses of a spinning black hole

Let us see whether by this purely mental analysis we can get to a finer point

(*The Adventure of the Three Gables*)

Given the extreme physical conditions that apply interior to its event horizon, it hardly seems wise to choose a black hole as a building platform or engineering tool. For those brave enough and with sufficient engineering savvy, however, there is much that might be gained by working within the region close to a rotating black hole. Indeed, one can in principle extract enormous amounts of energy from a black hole. In this respect, two main mechanisms for energy generation have been explored. The first is that of the Penrose mechanism, as outlined by Roger Penrose in 1969 [23], and the second is that of the Blandford–Znajek mechanism, as described by Roger Blandford and Roman Znajek in 1977 [24].

The Penrose mechanism relies on the general relativistic process of frame-dragging which operates close to, but outside of the event horizon, of a spinning black hole. In this sense, spacetime is literally dragged around by the black hole, opening-up a region called the ergosphere (Figure 4.5). Accordingly, by allowing an object to fall towards the boundary of the ergosphere about a rotating black hole, and letting it eject a fraction of its mass directly into the black hole at the moment of closest passage, the remaining mass will exit the region of the ergosphere with more energy than it had prior to its entry. Effectively, the ejected component gains energy at the expense of the black hole's spin. This is not a perpetual motion energy source, however, since sooner or later, the black hole will be spun down, and then the condition of frame-dragging will no longer apply. In principle, however, the Penrose Mechanism offers a way of

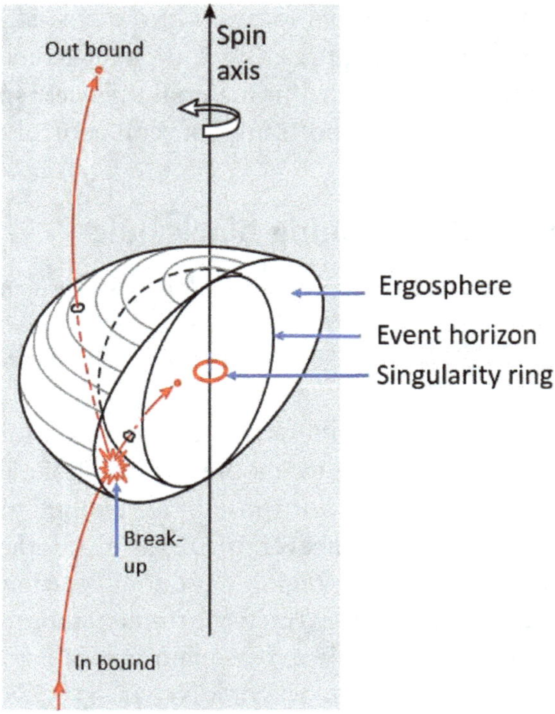

Figure 4.5. The ergosphere about a spinning black hole. The steps in the Penrose process are illustrated by the red path. The inbound object breaks apart in the ergosphere, with some material crossing over the event horizon. The outbound (remnant) fragment will have more total energy than that of the inbound (pre-breakup) object. Image adapted from Ruffini and Wheeler [25].

increasing the kinetic energy of an object set on a special orbit that just penetrates the ergosphere of a rotating black hole. It is the excess energy of the outgoing fragment that is harvestable.

The Blandford–Znajek mechanism relies upon the presence of an accretion disk about a rotating black hole. Specifically, if this disk supports a strong magnetic field, and if the field lines can thread into the ergosphere, then they will become twisted around the spin axis of the black hole. Any charged particles within the cylinder of the twisted magnetic field lines will then be accelerated away from the black hole in the form of jets — one below the disk, one above. As a result of this acceleration, the charged particles will emit synchrotron radiation, and this radiation might (at least in principle) be harvested and utilized by a suitably advanced civilization. Once again, however, the key driving mechanism is the spin energy of the rotating black hole. A recent variant on these energy extraction mechanisms has been developed by Luca Comisso (Columbia University) and Felipe Asenjo (Universidad Adolfo Ibáñez, Santiago) [26], who have investigated the potential effects of magnetic field line reconnection events within the ergosphere. Accordingly, they show that such events can accelerate particles to negative energies and that (in principle) large amounts of energy can be extracted from the black hole. Comisso has further suggested that, "thousands or millions of years from now, humanity might be able to survive around a black hole without harnessing energy from stars," adding that, "it is essentially a technological problem. If we look at the physics, there is nothing that prevents it." There is, of course, one problem with this scenario, in the case of our distant descendants at least, and that is the actual finding of a rotating black hole about which to begin engineering. Potentially, of course, more advanced alien civilizations might be more fortunate with respect to a conveniently placed black hole. Pushing our imaginations to the edge of incredulity, however, we might also speculate about the possibility of our deep descendants and/or an advanced alien civilization manufacturing a tailor-made black hole for the sole purpose of living off its energy.

The idea of creating a custom-made black hole is not entirely mad — recall the opening part of Section 4.4. Such ideas do, however, push us to the very boundaries of engineering extremes, and verge upon the miraculous. Nonetheless, Avi Loeb (Center for Astrophysics, Harvard–Smithsonian Observatory) recently speculated [27] that should we, or an alien civilization, find a way to engineer even the smallest possible mass black hole, with a mass of some 10^8 kilograms (recall footnote 13 in Section 4.4), then, in principle, energy could be extracted from it for millennia on end. At this mass, we are dealing with a black hole having an event horizon that is barely more than a few Planck lengths across — a mini black hole

(MBH), to say the least, and one easily lost if not careful confined. Such MBHs, if produced in the formative big bang of our Universe, will no longer exist, and so they would have to be manufactured. The reason why such small mass MBHs are no longer found in our Universe was explained by Stephen Hawking in his now classic and rightly famous research paper published in 1971. Quantum processes working at the event horizon of a black hole, Hawking argued, result in the emission of radiation (Hawking radiation) into space, and the slow evaporation (mass reduction) of the black hole itself. The evaporation time T_{evap} against Hawking radiation was discussed earlier (in Section 4.4), and for a 10^8-kg MBH, the lifetime against 'evaporation' is about 10 years. Clearly, such a lifetime is very, very short compared to the age of the Universe (short even by human standards), but as Loeb points out, one could feed the MBH so that its mass remains constant. The exact mechanism for such feeding is far from obvious, the event horizon of such an MBH being very much smaller than even the classical size of an electron (about 5.6×10^{-15} meters across). However, if one can make the MBH accrete just 2.5 kg of material per year, then this will counter the mass loss resulting from Hawking radiation.

Why would one want to go through all this effort is, perhaps, the next question. The point being, as Loeb emphasises, that black holes, as a result of Hawking radiation, have a luminosity L_{HR} and a temperature T_{HR}, and these quantities have incredibly high values at low masses. Indeed, the temperature depends upon the inverse mass of the black hole and is given by $T_{HR} = 1.2 \times 10^{15}$ $(M / 10^8 \text{ kg})^{-1}$ Kelvin. The luminosity depends upon the inverse mass squared, with $(L_{HR} / L_{\odot}) = 9.3 \times 10^{-11}$ $(M / 10^8 \text{ kg})^{-2}$. Placing a 10^8-kg MBH in a circular orbit (of altitude h) about a planet will then provide a surface energy flux $F_{MBH} = L_{HR}/4\pi h^2 = 2.96 \times 10^9 / (h(\text{km}))^2$ W/m^2. This flux would be equivalent to the solar constant of 1,361 W/m^2 for an orbital altitude of h(km) = 1,475. The very high temperature of 10^{15} K for the Hawking temperature of our 10^8-kg MBH dictates that the peak of the radiation will be in the far gamma ray part of the electromagnetic spectrum, with Wien's law (recall Section 1.5) giving $\lambda_{max} \approx 3 \times 10^{-18}$ m. Photons with this peak wavelength (or peak frequency of

10^{26} Hz) will have an incredible high penetrating power, and carry a hefty energy punch of some $E_{max} \approx 6.6 \times 10^{-8}$ Joules. The radiation dosage required to kill a human being, with a 50% probability, in 30 days is 4–5 Sievert (Sv), where 1 Sv = 1 Joule/kg. For an unshielded, 100-kg person, with a cross-sectional area of 1 m^2, a fatal dosage of 5 Sv will be accumulated within about 1/3 of a second. When extracting energy from an MBH, one is indeed playing with a deadly fire, but living deep underneath a planet's surface would provide some protection against the extreme radiation conditions that might prevail on the surface.

Well, we have made an idealized calculation of the possible energy that might be available from a radiating MBH. But, of course, several (major) problem steps have yet to be addressed. The radiation danger from such an object is one thing, the manufacture of it is entirely another, and how such a minuscular body might be entrained, and made to accrete material is far from clear. Certainly, at the present time, the generation process is completely beyond human capabilities. And, more fundamentally, it still has to be established that structures like MBHs can actually form, or have any real existence beyond the formulae of theoretical physics. Could, however, any of these existential construction and engineering uncertainties ever change? To which the only answer is: who knows? Loeb appreciates these points, and presents us with at least a testable idea, suggesting that, perhaps, such an orbiting MBH could be used to generate power for an advanced civilization located upon a rogue planet. Such rogue planets (that is, planets not in orbit about any specific star) are known to exist, and indeed, there might well be a vast number of such objects within our Milky Way galaxy. If such a situation is playing out, however, as Avi Loeb notes, it would appear to a terrestrial observer as a decidedly strange astrophysical object. Not only would an intense gamma ray source be associated with an otherwise inert planet, the gamma ray source would undergo periodic modulation, blinking out every time it is occluded by the planet's body in our line of sight. For a terrestrial mass rouge planet, with an artificial MBH moon, on a circular orbit of altitude 1,500 km, modulation would take place at intervals of about one hour. Pushing his idea to the extremes of credulity, however, Loeb concludes his research paper with an inflammatory suggestion, "Currently, there are a large number of startup companies aiming to make compact fusion reactors. A mini black hole would be far more efficient and environmentally-friendly." [26] Loeb's comments are arguably true (except for the MBH moon being environmentally-friendly bit), but fusion reactors are within reach of human construction in the here and now[20]. An ability to create even the smallest-possible-mass black hole is nowhere near feasible at the present, and/or even in the far distant future.

4.6. The Elysium Fields

Instead of Niagara dwarfing the human mind, the human mind can very readily dwarf Niagara.

(Arthur Conan Doyle — *The Narrative of John Smith*)

As I write this chapter, the Arecibo message has travelled some 50 light years from the solar system. Transmitted from the Arecibo Radio Telescope on 16 November 1974, the message, composed of 1,679 binary digits, is intended to tell the would-be receiver about us and where we are from. Within the volume of space with radius 50 light years (15 pc) about the Sun, there are something like 1,400 stars. Not all of these stars will be exposed to the Arecibo message, since it was a beamed signal, and, in fact, directed towards the globular cluster M 13, located about 25,000 light years away. The Arecibo message has to date completed just 0.2% of its journey. The selection of M 13 as a target for the signal was pure happenstance, and adopted simply because it was conveniently placed in the sky at the time that the message was scheduled to be sent out. In times since the 1970s, SETI programs have tended to consider globular clusters as unsuitable sites for study. The main reason for this being the low metallicity of their constituent Population II stars.

[20] There is, of course, the long-running joke, that physicists and engineers have been saying for the last 65 years that practical fusion power is just ten years away.

Since planets are composed mostly of elements that are metals (recall, in astronomy speak, metals are all those elements other than hydrogen and helium), it is reasoned that planets are unlikely to exist within globular clusters. Assuming that a planet is needed in order for life to develop, it has accordingly been argued that advanced civilizations will not exist within globular clusters. But, is this really the case? An advanced civilization may not have its genesis within a globular cluster, but it might reasonably thrive there [28].

In terms of planets within globular clusters, at the present time only one such example has been found, this being a 2.5 Jupiter mass planet in orbit about a binary system, composed of a white dwarf and a neutron star, within NGC 6121 (Messier 4). Indeed, the neutron star is a millisecond pulsar (PSR B1620–26 A) spinning upon its axis some 90 times per second. The planet has been grandly named *Methuselah* and 'the Genesis planet', and with an estimated age of 12.5 billion years, it is the oldest known planet [29]. This particular planet orbits the central system at a distance of some 23 AU, with an orbital period of about 100 years. It is not presently clear if *Methuselah* formed at the same time as the host binary system, or whether it formed from the gas and dust debris ejected during the planetary nebula and supernova phases associated with the host stars. In general, exoplanet statistics indicate that the lower the metallicity of a star, the less likely it is that that it will have a Jovian planetary companion. This rule, however, may not apply to terrestrial planets, and while extensive surveys of the globular cluster 47 Tucanae (recall Figure 3.15) have revealed no terrestrial exoplanet candidates, this is not the same as saying that none actually exist there. The main reason why it is generally thought that planets are unlikely to form and/or survive in orbit about a specific natal star is entirely due to the close star packing that exists within globular clusters. To a sufficiently advanced civilization, however, whether a planet is in orbit about a star or not is not necessarily a critical issue. The planet is a resource to them rather than a necessity.

It was argued earlier (Section 4.1.6) that the existence of the Fermi Paradox suggests that long-term survival on artificial space habitats has not been a successful means of prolonging civilization survival. This argument can be modified, however, to include the possibility that rather than setting themselves adrift within the disk of the Milky Way galaxy, once their nurturing star has evolved off the main sequence, an advanced civilization might deliberately migrate towards their nearest globular cluster. Indeed, for an advanced civilization, a globular cluster is full of starlight (from which energy can be extracted) and raw materials — the raw materials in this case being an assumed population of free-floating planets and the stars themselves. An explanation for the Fermi Paradox now becomes: they (that is, advanced aliens) are not visiting us here in the solar system, located as it is within the disk of the Milky Way galaxy, because they have migrated to the globular clusters in the outer regions of the galaxy, and out of the galactic plane. In these outer reaches, and especially within globular clusters, stellar change runs much more slowly, and potential home-stars are long-lived and plentiful.

There are distinct advantages in moving to a globular cluster. Since the stars are so closely packed — the average separation between stars being in the order of 0.1 pc (or about 1 light year) — communication time between host systems will be on the timescale of months to years, and travel time[21] between systems will be on a timescale of decades to centuries. By having multiple centers of settlement, existential risks are minimized. Furthermore, free-floating planets (should they exist) can be mined for resources, and activities such as star-lifting (Section 4.1.3) and/or star-on-star collision manipulation are made more straightforward by the short separation distances. Indeed, by engineering collisions within a globular cluster, an advanced civilization could in principle generate main sequence stars (that is, objects that will look to us like blue stragglers), and they might potentially attempt to form stellar-mass black holes by arranging material to be gradually accreted onto a neutron stars. Indeed, to a Kardashev II advanced civilization, a globular cluster is an ideal environment in which to both live and thrive. Are all blue

---

[21]Traveling at say 1/100 the speed of light, the time to move from one system to another ($d \sim 1$ light year) will be in the order of 100 years.

stragglers in globular clusters formed artificially? No, certainly not, but if one can eliminate a formation path according to the natural process outlined in Chapter 3 (recall Figure 3.1), then, in Sherlockian terms, what remains, no matter how bizarre, could potentially be the indicator of an advanced civilization at work. This raises the distinct possibility that, in equation (3.8), $N_{\text{other}} > 0$.

The possibilities for energy generation, by a technically advanced civilization, within a globular cluster are manyfold. There is plenty of starlight, there is the potential for energy generation from black holes via the Penrose process, and as argued for by Clément Vidal [30], there are also numerous binary stars to tap. Indeed, Vidal argues for the potential presence of starivores within globular clusters. Such civilizations are defined according to their use of stellar energy tapped from within the confines of an accreting binary system, with the dense primary being a white dwarf, neutron star, or black hole. Such civilizations will not live on a planet, but at the L_1 location of the system (recall Figure 3.6), extracting energy from the matter being transferred to the compact companion. Indeed, such starivore civilizations might attempt to manipulate the flow rate of matter transfer, thereby signalling their existence to the rest of the galaxy. Here, the SETI signal might be detected according to modulations within the X-ray emissions and/or gamma ray outbursts from compact binary systems.

4.7. The search ever continues

And yet it was not entirely a question of imagination

(*The Hound of the Baskervilles*)

It is a non-trivial exercise, fraught with uncertainties, to seek-out the signs of an extraterrestrial civilization. We do not know *a priori* where to look, what to look for, and indeed what to even expect. The search, however, is important for many reasons, not least that of galactic companionship. Knowing of the existence of extraterrestrial civilizations is one thing, but communicating with them is another. If it is found that the most advanced civilizations have migrated to the outer boundaries of the galaxy and/or specifically to the globular clusters, then any chance of our physically knowing them is lost to us entirely. This is perhaps as it should be, and, indeed, rather than hoping to receive the benevolent wisdom of others from afar, our long-term goal should really be to know ourselves, and from that knowledge and understanding prosper, and then reach, in a physical sense, for the stars. For all that, amongst the blue stragglers, the answer to our search for even a faint sign of companionship may be writ large on the sky.

In the mind's eye, miracles and marvels can be readily realized, and it is easy enough to spin words of wonder together. For all that, nothing in this chapter is based upon any assumptions or concepts that are beyond the realm of known physical principles at the present time. That is to say, while we might not currently know how to directly perform the operations, or enact the required engineering, or know how to make the necessary material structures, none of them specifically require the invocation of processes operating beyond the rules of fundamental chemistry, industry, and physics. This by no means indicates that we, the inhabitants of Earth, at the present time, or even our very distant descendants, will be able to develop such technologies and turn the Sun into a blue straggler. Likewise, it does not mean that any galactic civilization, highly advanced or otherwise, will be able to astero-engineer blue stragglers either, but it does give us abundant food for thought about possibilities. And if life and civilization-building is anything, it is all about free-thinking and the opening-up of future possibilities under the constraint of humility.

> *I have never been nervous about details, and one must be masterful sometimes*
>
> (Arthur Conan Doyle)

References

[1] M. Beech. *Introducing the Stars: formation, structure and evolution.* Springer Nature, Switzerland AG. 2019.

[2] M. Beech. *Rejuvenating the Sun and avoiding other global catastrophes*. Springer, New York, 2008.

[3] See, e.g., R. Kippenhahn, A. Weigert, and A. Weiss. *Stellar Structure and Evolution*. Springer, Berlin, 2012.

[4] A. Paris, and E. Davies. Hydrogen Clouds from Comets 266/P Christensen and P/2008 Y2 (Gibbs) are Candidates for the Source of the 1977 "WOW" Signal. *Washington Academy of Sciences*, 2015. See: https://arxiv.org/pdf/1706.04642.

[5] M. Scoggins, and D. Kipping. Lazarus stars: numerical investigations of stellar evolution with star-lifting as a life extension strategy. *Monthly Notices of the Royal Astronomical Society*, **523**, 3251–3257, 2023.

[6] M. Fogg. Solar exchange as a means of ensuring the long-term habitability of the Earth. *Speculations in Science and Technology*, **12**, 153–157, 1989.

[7] A. Stevens, D. Forgan., and J. James. Observational signatures of self-destructive civilizations. *International Journal of Astrobiology*, **15**, 333–344, 2016. https://doi:10.1017/S1473550415000397.

[8] J. Greaves, *et al*. Phosphine gas in the cloud decks of Venus. *Nature Astronomy*, **5**, 655–664, 2021.

[9] M. Beech. Terraformed exoplanets and SETI. *Journal of the British Interplanetary Society*, **60**, 43–46, 2007.

[10] Project Hephaistos. https://www.astro.uu.se/~ez/hephaistos/hephaistos.html.

[11] B. Carter. Five- or six-step scenario for evolution? *International Journal of Astrobiology*, 7, 177–182, 2008. https://doi.org/10.1017/S1473550408004023

[12] D. Mills, J. Macalady, A. Frank, and J. Wright. A reassessment of the "hard steps" model for the evolution of intelligence. *Science Advances*, 11, eads5698, 2025.

[13] M. Beech. Aspects of an astero-engineering option. *Journal of the British Interplanetary Society*, **46**, 317–322, 1993.

[14] M. Beech. Blue Stragglers as indicators of extraterrestrial civilizations? *Earth, Moon and Planets*, **49**, 177–186, 1990.

[15] M. Fogg. Stellifying Jupiter: A first step to terraforming the Galilean satellites. *Journal of the British Interplanetary Society*, **42**, 587–592, 1989.

[16] M. Beech. A dark sun rising: it's a solar wrap. *Journal of the British Interplanetary Society*, **63**, 104–107, 2010.

[17] See, https://vascoproject.org/vanishing-stars/.
[18] See: https://photojournal.jpl.nasa.gov/catalog/PIA21467.
[19] M. Beech. *The Large Hadron Collider: unravelling the mysteries of the universe*. Springer, New York, 2010.
[20] S. Hawking. Gravitationally collapsed objects of very low mass. *Monthly Notices of the Royal Astronomical Society*, **152**, 75–78, 1971.
[21] K. Thorne, and A. Żytcow. Stars with degenerate neutron cores. I — Structure of equilibrium models. *The Astrophysical Journal*, **212**, 832–858, 1977.
[22] E. Bellinger, *et al.* Solar evolution models with a central black hole. *The Astrophysical Journal*, **959**, A113, 2023.
[23] R. Penrose, and R. Floyd. Extraction of rotational energy from a black hole. *Nature Physical Sciences*, **229**, 177–179, 1971. https://doi.org/10.1038/physci229177a0.
[24] R. Blandford, and R. Znajek. Electromagnetic extraction of energy from Kerr black holes. *Monthly Notices of the Royal Astronomical Society*, **179**, 433–456, 1977. https://doi.org/10.1093/mnras/179.3.433.
[25] Ruffini, R., and Wheeler, J. A. Introducing the Black Hole. *Physics Today*, April, 2009.
[26] See: https://news.columbia.edu/news/could-we-harness-energy-black-holes.
[27] Loeb, A. Illumination of a Planet by a Black Hole Moon as a Technological Signature. *Research Notes of the American Astronomical Society*, 8(3), 2024.
[28] R. Di Stefano, and A. Ray. Globular clusters as cradles of life and advanced civilizations. *The Astrophysical Journal*, **827**, A54, 2016.
[29] D. Backer, R. Foster, and S. Sallman. A second companion to millisecond pulsar 1620–26. *Nature*, **365**, 817–819, 1993.
[30] Vidal, C. *The Beginning and the End: the meaning of life in a cosmological perspective*. Springer, New York, 2014.

Afterword

Swung out of sunlight into cosmic shade,
Come what come may the imagination's heart
Is constellation high and can't be weighed.
(Mervyn Peake — *To live at all is miracle enough*)

Seeking answers to the blue straggler mystery has led us on a merry dance — and yet, the music and motion continues. Indeed, if the history of science tells us anything, it tells us that the music never really ends. We have a beginning, a middle, and a strong hint of the ending to our blue straggler mystery, but we also have lingering questions and loose ends that have yet to be tidied away. We have covered much exploration, observation, history, thinking, speculation, time, space, computation, and false starts, but the full story of the blue stragglers has yet to be fully writ large. We began our investigations with two chapters on how astronomers have labored to understand the stars and, as it were, identify the usual suspects at play in the day-to-day celestial realm. Indeed, astronomers have developed a vast array of subtle tools and methodologies for teasing information out of starlight and stellar motions. That we know the stars, their size, mass, temperature, luminosity, space motion, and orbits is certainly one of the greatest achievements of our modern age; it is testament to past, present and (hopefully) future human intellect and tenacity.

It was from the hard-won, normalized background of typical star properties that the blue stragglers, as a real phenomenon, emerged — they were different and did not fit in with the greater crowd. Indeed, the blue stragglers are different — different, but as we have seen explainable according to what are now standard principles applied in ways that were undreamed of a century, even half-a-century, ago. By coming to grips with the pathways and circumstances under which blue stragglers might form, a new branch on the astrophysical tree of knowledge has been delineated. What was lacking in the early days of the investigation was an understanding of how compact binary systems might evolve and a limited understanding of extreme stellar evolution when stars collide — these areas of study not opening up until effectively the close of the 20th century. And, when we say a limited understanding, this is not to say that there was a lack of intellect, or a lack of daring, or a lack of imagination, but a limit to what could be computationally modeled. Pen, paper, human tenacity, and mathematical simplifications could identify the likely directions towards which solutions to the blue straggler mystery might lie, but it was the advancement in detailed computational ability, the bridging of a technology gap, that needed to come about before a clear path forward became available.

Is the blue straggler mystery fully explained and solved? No. Not a bit of it, but it has certainly been changed and reduced to a much better level of understanding. The deeper mystery is perhaps no more, but this does not mean that we fully understand all the ifs, ands, and buts of blue straggler formation. This situation illustrates both the power, as well as the intrigue, of scientific progress. Indeed, it is from Lord Alfred Tennyson that we learn, "science moves, but slowly, slowly, creeping from point to point." Given that the blue straggler mystery is now largely understood, its full resolution points us towards new areas of thought, observation, experimentation, and computer modeling, and it points towards the need to consider ever more complex, ever more extreme, and ever deeper mysteries. Astronomy, as an active and vibrant science, is alive and doing very well, thank you very much.

There is a hypothetical question that we must now ask. Would Sherlock Holmes be satisfied with the progress that astronomers have made in unraveling the blue straggler mystery? The answer to this, I would wager, is both yes and no. While he would acknowledge that the main points have been resolved, he would still worry about all the loose ends. This, of course, is entirely due to the fact that Holmes generally expected to make a watertight and indisputable case against his combatants. Scientists, as indicated in the Introduction, rarely have the satisfaction of such definite closure. Indeed, even Holmes would have to admit in the *Adventure of the Final Problem*, that in spite of his detailed plans and heroic actions, Professor Moriarty still managed to escape the net placed around his criminal empire[1]. When it comes to the continuing saga of the blue stragglers, there will, I have no doubt, be future surprises, unanticipated shocks, and new revelations. Above all else, we simply need to keep looking (Figure A1). Indeed, no cosmic mystery is ever really

Figure A1. Holmes the investigator. "For a long time he remained there, turning over leaves and dried sticks ... examining with his lens not only the ground, but even the bark of the trees as far as he could reach." *The Boscombe Valley Mystery.* Drawing by Sidney Paget. Image from https://www.arthur-conan-doyle.com/images/6/63/Illus-bosc-paget-07.jpg.

[1]So too did Colonel Sebastian Moran, "the second most dangerous man in London", escape Holmes' net.

solved; it simply moves away from its original starting point and reappears elsewhere in a Universe that is truly devious in its design.

While this book is intended to primarily act as an introduction to and a rudimentary explanation of the blue straggler mystery, I have also dared to dream within its pages. Dreaming, of course, is a vital (but rarely acknowledged) part of the scientific process, and history tells us that scientific dreams can inspire the waking hours of the human mind. In this respect, I have dared to think beyond the norms, and I have asked how and why an advanced alien civilization might engineer a star into a long-lived, extended–main-sequence, blue-straggler state. And in doing this, I have attempted to stay on the science, rather than the wild–pipe-dream side of physical reality. Indeed, taking a glance at the poetic lines that open this Afterword, I think Peake strikes to the core of what the pursuit of science is largely about, and indeed, the imagination's heart will only sing if it is allowed to fly unrestrained and constellation-high. For all that, if there is one thing better than an astronomical mystery, it is the solution to an astronomical mystery, and astronomers, perhaps more than other scientists and artists, have a vast canvas upon which to direct their thoughts and find their muse. Speculation, even rampant speculation, is a vital part of the scientific process, and it is from the spun-out threads of tempered scientific dreaming that new ways of looking at and understanding the heavens come about. To live at all is miracle enough, and we live in an incredibly complicated Universe that only slowly shares its deeper secrets. And, who knows, perhaps our very, very distant descendants might attempt to turn the Sun into a Lazarus star (recall Chapter 4.1.3), and engineer their very own solar system blue straggler.

Now nearly one-and-a-half centuries ago, the Victorian practitioners of science had a somewhat naïve notion of scientific advancement, thinking that it proceeded in a linear fashion like some stately river, emerging from some distant well-head, and thereafter pushing every forward into new territories of understanding towards a distant sea. Of course, the actual *doing* of science and the act of scientific discovery is nowhere near as straightforward as the mere flow and accumulation of new data. Indeed, the river of knowledge is far removed from an ambling Thames, or kingly Nile. No, it is a watercourse with many hidden and obscure tributaries, rapids, weirs, large pooling lochs, and reflective lakes. Its river banks can sometimes be high and absolutely constraining; they can sometimes be low, and allow for a veritable flood-plain to open out. Isaac Newton famously suggested, with feigned modesty, that he had just picked up a few sparkling pebbles from along the shores of a vast ocean of knowledge. And, indeed, the various river courses of knowledge that humanity has actually found — and even, in some cases, successfully navigated — all empty into a vast sea of understanding. But there are still unknown rivers, rills, and tributaries that have yet to be found and mapped. And, furthermore, if there is one thing that we do know about standing upon the shores of a vast lake or ocean, it is that there is a very definite distant horizon, and that the horizon itself pulls at our imagination. Indeed, the most primordial of human responses to any horizon is the irresistible urge to see what lies beyond it. What lies beyond the solution horizon of the blue straggler mystery, is, no doubt, yet more mystery — and that is a pleasing thought and an encouraging place to end this particular dialog.

> *Chaos, Cosmos! Cosmos, Chaos! who can tell how all will end!*
> *Read the wide world's annals, you, and take their wisdom for your friend.*
>
> (A. L. Tennyson — *Locksley Hall Sixty Years After*)

www.ingramcontent.com/pod-product-compliance
Lightning Source LLC
LaVergne TN
LVHW010334220926
848484LV00004BB/894
* 9 7 8 9 8 1 9 8 2 0 0 8 5 *